The ESSENTIALS of MECHANICS III

Staff of Research and Education Association,
Dr. M. Fogiel, Director

This book is a continuation of *"THE ESSENTIALS OF MECHANICS II"* and begins with Chapter 17. It covers the usual course outline for Mechanics III. *"THE ESSESTIALS OF MECHANICS I"* and *"THE ESSENTIALS OF MECHANICS II"* cover earlier topics.

Research and Education Association
505 Eighth Avenue
New York, N.Y. 10018

THE ESSENTIALS OF MECHANICS III

Printed in the United States of America

Library of Congress Catalog Card Number 87-61798

International Standard Book Number 0-87891-613-X

WHAT "THE ESSENTIALS" WILL DO FOR YOU

This book is a review and study guide. It is comprehensive and it is concise.

It helps in preparing for exams, in doing homework, and remains a handy reference source at all times.

It condenses the vast amount of detail characteristic of the subject matter and summarizes the **essentials** of the field.

It will thus save hours of study and preparation time.

The book provides quick access to the important facts, principles, theorems, concepts, and equations of the field.

Materials needed for exams, can be reviewed in summary form – eliminating the need to read and re-read many pages of textbook and class notes. The summaries will even tend to bring detail to mind that had been previously read or noted.

This "ESSENTIALS" book has been carefully prepared by educators and professionals and was subsequently reviewed by another group of editors to assure accuracy and maximum usefulness.

Dr. Max Fogiel
Program Director

CONTENTS

This book is a continuation of *"THE ESSENTIALS OF MECHANICS II"* and begins with Chapter 17. It covers the usual course outline for Mechanics III. *"THE ESSESTIALS OF MECHANICS I"* and *"THE ESSENTIALS OF MECHANICS II"* cover earlier topics.

Chapter No. **Page No.**

CHAPTER 17

PLANE MOTION OF RIGID BODIES

17.1 INTRODUCTION

Plane Motion - Rigid body motion where the particles composing the body are always a fixed distance from a reference plane.

Plane of Motion - The plane in which the center of mass moves.

General plane motion may be considered to be a sum of a translation and a rotation.

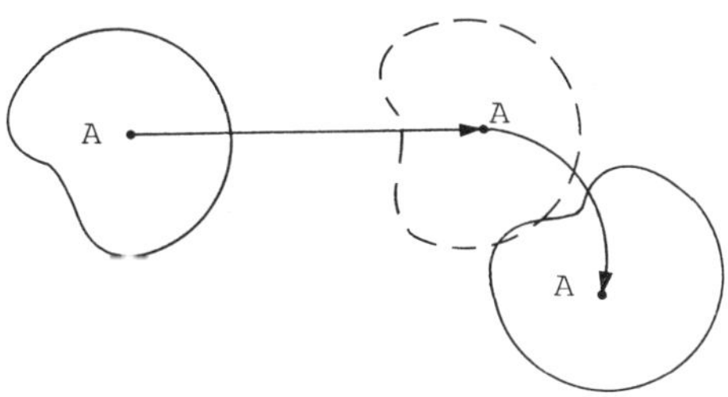

Fig. 17.1 General Plane Motion

17.2 CENTER OF MASS

General Formulas:

$$x_{cm} = \frac{\int_Q PxdQ}{\int_Q PdQ} \; ; \quad y_{cm} = \frac{\int_Q PydQ}{\int_Q PdQ} \; ; \quad z_{cm} = \frac{\int_Q PzdQ}{\int_Q PdQ} \tag{17.1}$$

For a Rigid Body:

Substitute into equation (17.1):

$P = \rho$ = Density of the Material

$Q = V$ = Volume

For a Thin Shell:

Substitute:

$P = m$ = Mass Per Unit Area

$Q = s$ = Surface Area

For a Thin Wire:

Substitute:

$P = n$ = Mass Per Unit Length

$Q = \ell$ = Length

For a Composite Body:

$$x_{cm} = \frac{x_1 m_1 + x_2 m_2 + \ldots}{m_1 + m_2 + \ldots} \tag{17.2}$$

$x_1, x_2 \ldots$ are center of mass locations for $m_1, m_2 \ldots$ respectively. y_{cm} and z_{cm} are calculated using equations similar to equation (17.2).

Symmetry Considerations - If the body has a plane or axis of symmetry, then the mass center is located on this plane or axis of symmetry.

A) Solid Hemisphere

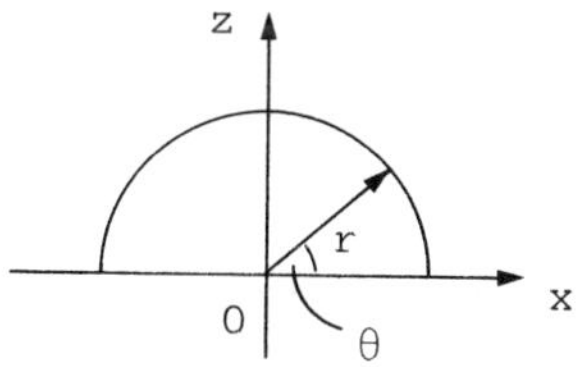

Fig. 17.2

$$z_{cm} = \frac{\int_0^r \rho\pi z(r^2-z^2)dz}{\int_0^r \rho\pi (r^2-z^2)dz} = \frac{3}{8} r \qquad (17.3)$$

$x_{cm} = y_{cm} = 0$

B) Hemispherical Shell

Refer to Figure 17.2:

$$z_{cm} = \frac{\int_0^r \rho\, 2\pi rz dz}{\int_0^r \rho\, 2\pi r dz} = \frac{1}{2} r \qquad (17.4)$$

$x_{cm} = y_{cm} = 0$

C) Semi-circular Arcs

$$z_{cm} = \frac{\int_0^\pi \rho\, (r\sin\theta) r d\theta}{\int_0^\pi \rho\, r d\theta} = \frac{2}{\pi} r \qquad (17.5)$$

$$x_{cm} = y_{cm} = 0$$

D) Semi-Circular Lamina

$$\boxed{z_{cm} = \frac{4}{3\pi} r} \qquad (17.6)$$

$$x_{cm} = y_{cm} = 0$$

17.3 STATIC EQUILIBRIUM

By Newton's Laws:

Translational Equilibrium:

$$\vec{F}_1 + \vec{F}_2 + \ldots = 0 \qquad (17.7)$$

Rotational Equilibrium:

$$(\vec{r}_1 \times \vec{F}_1) + (\vec{r}_2 \times \vec{F}_2) + \ldots = 0 \qquad (17.8)$$

For a particle in a uniform gravitation field:

Let $m\vec{g}$ = Gravitational Force Acting on the Particle

Translational Equilibrium:

$$\vec{F}_1 + \vec{F}_2 + \ldots + m\vec{g} = 0 \qquad (17.9)$$

Rotational Equilibrium: (17.10)

$$(\vec{r}_1 \times \vec{F}_1) + (\vec{r}_2 \times \vec{F}_2) + \ldots + (\vec{r}_{cm} \times m\vec{g}) = 0$$

17.4 EQUATIONS OF MOTION

Basic Equations:

$$\Sigma \vec{F} = m\vec{a}_{cm}$$

$$\Sigma \vec{M}_G = \dot{\vec{H}}_G = I_G \vec{\alpha} \qquad (17.11)$$

G = Mass Center

a_{cm} = Acceleration of Mass Center

α = Angular Acceleration

M_G = Moment About Mass Center

I_G = Moment of Inertia About G

$\dot{\vec{H}}_G$ = Angular Acceleration

Scalar Form of Equation (17.11):

$$\Sigma F_x = m(a_x)_{cm}; \quad \Sigma F_y = m(a_y)_{cm}; \quad \Sigma M_G = I_G \alpha \qquad (17.12)$$

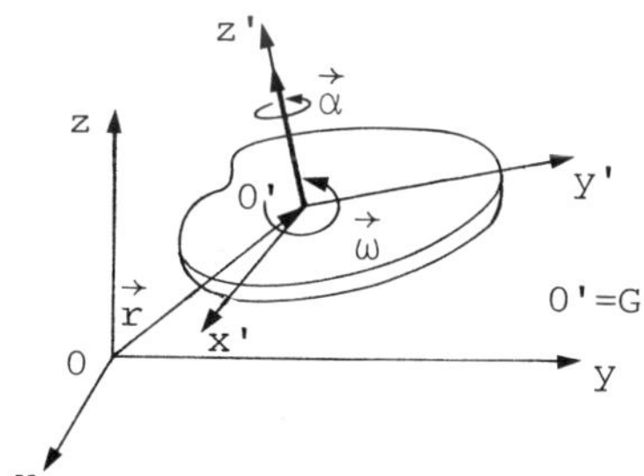

Fig. 17.3

17.5 ANGULAR MOMENTUM

The general theorem for angular momentum is applicable even for cases in which the axis of rotation is not fixed:

For each particle of the body:

$$\vec{r}_i = \vec{r}_{cm} + (\vec{r}_i)_{cm} \tag{17.13}$$

where

$(\vec{r}_i)_{cm}$ = Position vector of the ith particle relative to the mass center

Also

$$\vec{v}_i = \vec{v}_{cm} + (\vec{v}_i)_{cm} \tag{17.14}$$

Angular momentum is given by:

$$\boxed{\frac{d}{dt}\left[\sum_i (\vec{r}_i)_{cm} \times m_i(\vec{v}_i)_{cm}\right] = \frac{d\vec{H}_G}{dt} = \Sigma\vec{M}_G} \tag{17.15}$$

Equation (17.15) holds good even if the mass center is accelerating. When using another point as reference the mass center must be stationary in an inertial reference system.

Applications

A) Centroidal Rotation - The axis of rotation passes through the mass center.

Equations of Motion:

$$\boxed{\begin{aligned} \Sigma\vec{F}_x &= 0 \\ \Sigma\vec{F}_y &= 0 \\ \Sigma\vec{M}_G &= I_G\vec{\alpha} \end{aligned}} \tag{17.16}$$

B) Noncentroidal Rotation

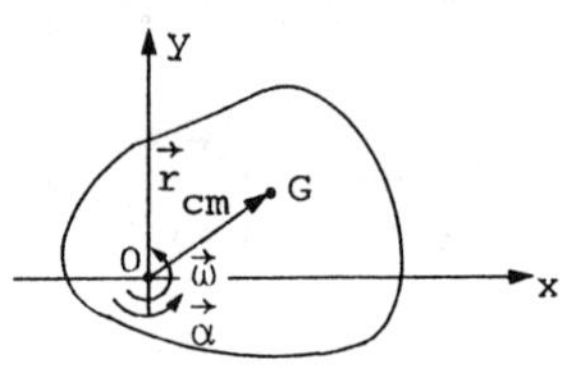

Fig. 17.4

$$\begin{aligned}\Sigma F_x &= m\, r_{cm}\, \omega^2 \\ \Sigma F_y &= m\, r_{cm}\, \alpha \\ \Sigma M_G &= I_G \alpha \quad \text{or} \quad \Sigma M_0 = I_0 \alpha\end{aligned} \tag{17.17}$$

C) Rolling Bodies

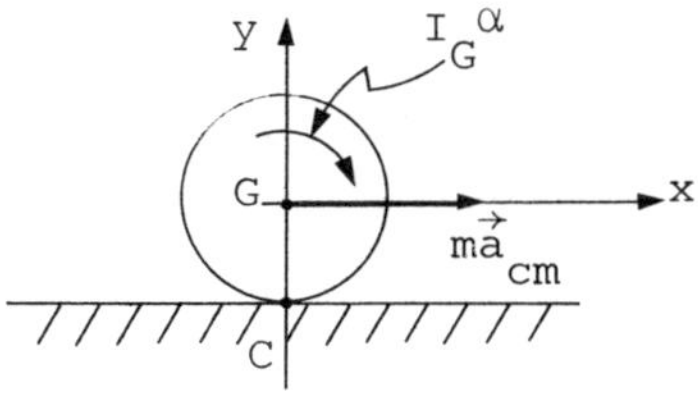

Fig. 17.5

Equations of Motion:

$$\begin{aligned}\Sigma \vec{F}_x &= m\vec{a}_{cm} \\ \Sigma \vec{F}_y &= 0 \\ \Sigma \vec{M}_G &= I_G \vec{\alpha}\end{aligned} \tag{17.18}$$

For rolling without slipping:

$$\Sigma M_c = I_c \alpha \tag{17.19}$$

c = Instant Center of Rotation

If the wheel is unbalanced:

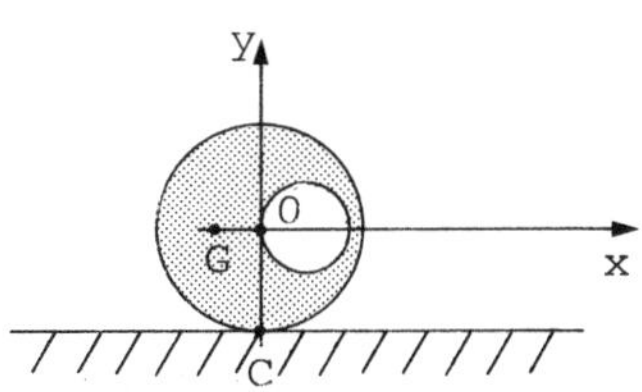

Fig. 17.6

Equation (17.19) is valid at points where the mass center G crosses the line connecting the geometric center O and the instantaneous center C.

17.6 D'ALEMBERT'S PRINCIPLE

D'Alembert's Principle - The resultant of the external forces applied to a body composed of a system of particles is equal to the vector sum of the effective forces acting on all particles.

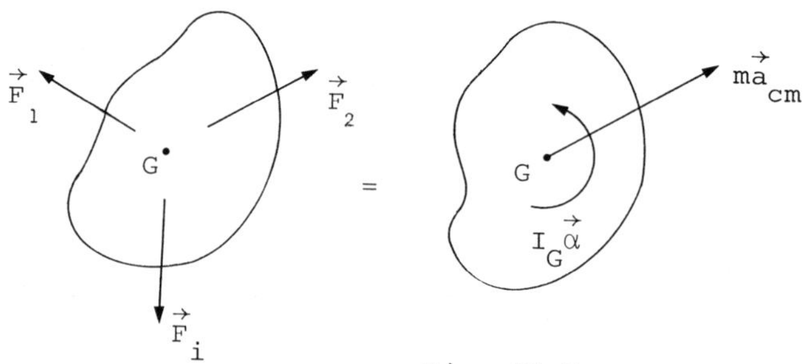

Fig. 17.7

Particular Cases:

A) Translation: $\vec{\alpha} \equiv 0$

Effective forces reduce to a vector $m\vec{a}_{cm}$ acting at the mass center G.

B) Centroidal Rotation: $\vec{a}_{cm} \equiv 0$ as axis of rotation passes through mass center G.

Effective forces reduce to $I_G\vec{\alpha}$ acting at G.

Note:

A) The motion of the mass center takes place as if the total mass and external forces are concentrated at that point.

B) In general, the external forces do not pass through the mass center.

C) The rigid body rotates about the mass center as if it was a fixed point, if the body is in plane motion.

17.7 SYSTEMS OF RIGID BODIES

For a system consisting of several rigid bodies, the method described in the preceding section may also be applied.

Method of Solution:

A) Draw a diagram of the system, including all the external forces and moments acting on the system.

B) Express the external forces as equipollent to a system of effective forces and solve for the unknown quantities.

C) This approach is valid for problems with no more than three unknowns.

17.8 CONSTRAINED MOTION

For rigid bodies moving under constraints, equations (17.11) may be applied to the mass center of the body.

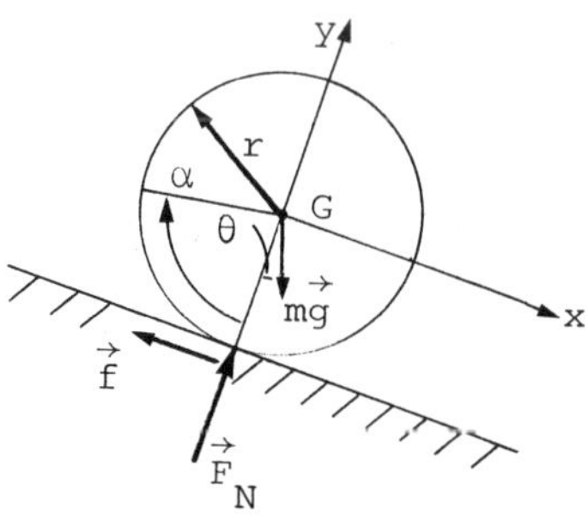

Fig. 17.8

For a body rolling down an incline:

Component equations of translation of the mass center:

$$m\ddot{x}_{cm} = mg\sin\theta - f$$
$$m\ddot{y}_{cm} = -mg\cos\theta + F_N \qquad (17.20)$$

From the constraints of the problem:

$$\begin{aligned} y_{cm} &= \text{constant} \\ F_n &= mg\cos\theta \end{aligned} \qquad (17.21)$$

The only force producing a moment is the frictional force $\vec{f}$. Therefore, the equation of rotation is:

$$I_g \dot{\omega} = fr \qquad (17.22)$$

Special Cases

A) No Slipping:

$$x_{cm} = r\psi \qquad (17.23)$$

where ψ is the angle of rotation and equation (17.22) becomes:

$$\frac{I_G}{r^2}\ddot{x}_{cm} = f \qquad (17.24)$$

Substituting and solving gives:

$$\boxed{\ddot{x}_{cm} = \frac{g\sin\theta}{1 + (k_{cm}^2/r^2)}} \qquad (17.25)$$

where k_{cm} = Radius of Gyration of G

Therefore, the body moves with constant linear and angular acceleration.

Angular Acceleration

Energy Equation:

$$\boxed{\frac{1}{2}m\dot{x}_{cm}^2 + \frac{1}{2}mk_{cm}^2\frac{\dot{x}_{cm}^2}{r^2} - mgx_{cm}\sin\theta = E} \qquad (17.26)$$

B) With Slipping:

Equation of Translation:

$$m\ddot{x}_{cm} = mg\sin\theta - \mu mg\cos\theta \qquad (17.27)$$

μ = Coefficient of Sliding Friction

Equation of Rotation:

$$I_G\dot{\omega} = \mu mg\, r\cos\theta \qquad (17.28)$$

If G has constant acceleration:

$$\ddot{x}_{cm} = g(\sin\theta - \mu\cos\theta) \qquad (17.29)$$

and if angular acceleration is also constant:

$$\dot{\omega} = \frac{\mu gr\cos\theta}{k_{cm}^2} \quad (\because\ I_G = mk_{cm}^2) \qquad (17.30)$$

By integration:

$$\dot{x}_{cm} = Ar\omega$$

where

$$A = \frac{k_{cm}^2}{r^2}\left(\frac{\tan\theta}{\mu} - 1\right) \qquad (17.31)$$

CHAPTER 18

ENERGY AND MOMENTUM METHODS - RIGID BODIES

18.1 PRINCIPLE OF WORK AND ENERGY

The work-energy method is an alternate approach to solving problems in dynamics. Its advantages are:

A) Usually faster and simpler to use.

B) Involves only scalar quantities.

C) Can be applied directly to a system without analysis of its component parts.

D) Directly relates forces, displacements and velocities.

Translation

$$\boxed{W = \frac{1}{2} m(v^2 - v_0^2) = \Delta kE} \qquad (18.1)$$

W = Work Done

v_0 = Initial Velocity

V = Final Velocity

KE = Kinetic Energy; units: ft-lb

m = Mass

Fixed Axis Rotation

$$\int_{\theta_0}^{\theta} \Sigma M d\theta = \frac{1}{2} I(\omega^2 - \omega_0^2) \qquad (18.2)$$

If ΣM is variable, integrate equation (18.2), or find the area under the M-Q diagram.

If ΣM is constant, then

$$\Sigma M(\theta) = \frac{1}{2} I(\omega^2 - \omega_0^2) \qquad (18.3)$$

The initial direction of rotation is considered to do positive work.

For rotation under gravity forces:

$$\int_{\theta_0}^{\theta} \Sigma M d\theta = Wh \qquad (18.4)$$

where h = Change in elevation of the mass center
$= \bar{r} \sin\theta$

For a system in translation and rotation:

$$W = \Sigma \frac{m}{2} (v^2 - v_0^2) + \Sigma \frac{I}{2} (\omega^2 - \omega_0^2) \qquad (18.5)$$

Work done by a force $\vec{F}$:

$$W = \int \vec{F} \cdot d\vec{r} = \int \vec{F}(dr\cos\theta) \qquad (18.6)$$

$d\vec{r}$ = Increment in the distance in the direction of the applied force

s = Path of the motion

θ = Angle between $\vec{F}$ and $d\vec{r}$

For constant forces:

$$W = \Sigma Fs = \text{Area under the force-displacement diagram} \quad (18.7)$$

Examples of Forces That Do No Work:

A) Reactions at a frictionless pin when the body rotates about the pin.

B) Reactions at a frictionless surface when the body moves on the surface.

C) Weight of a body when the mass center moves horizontally.

D) No work is done by the frictional force at the point of contact, when a body rolls on a surface without slipping.

18.2 KINETIC ENERGY IN PLANE MOTION

Since plane motion is equal to translation plus rotation, kinetic energy is equal to the work equation as given by equation (18.5):

$$\Delta KE = W = \Sigma \frac{m}{2}(v^2 - v_0^2) + \Sigma \frac{I}{2}(\omega^2 - \omega_0^2) \quad (18.8)$$

In terms of the instant center of rotation:

$$\Delta KE = \frac{1}{2} I_c \omega^2 - \frac{1}{2} I_{c_0} \omega_0^2 \quad (18.9)$$

I_c, I_{c_0} = Moment of inertia about the final and initial positions of the instant center c, respectively

For a homogeneous free-rolling wheel, I_c is constant, and:

$$\Sigma M_c = \frac{1}{2} I_c(\omega^2 - \omega_0^2) \tag{18.10}$$

Note: The above equation is valid only when there is no slippage. If slippage occurs, then equation (18.8) should be used, i.e., work done by the kinetic friction forces should also be considered.

18.3 CONSERVATION OF ENERGY

If a rigid body, or a system of rigid bodies are in motion under the influence of conservative forces, then the equation of conservation of energy (14.37) for a particle holds good or rigid bodies also.

$$(KE)_1 + (PE)_1 = (KE)_2 + (PE)_2 = E \tag{14.37}$$

For Plane Motion

The kinetic energy term should account for both, translational and rotational effects.

18.4 POWER

Power is the time rate at which work is done.

$$\text{Power} = \frac{dW}{dt} \tag{18.11}$$

In terms of forces:

$$\text{Power} = \vec{F} \cdot \vec{v} \tag{18.12}$$

In terms of moments:

$$\boxed{\text{Power} = M\omega} \tag{18.13}$$

where

$$\omega = \frac{d\theta}{dt}$$

Efficiency is given by equation (14.9):

$$\boxed{\eta = \frac{\text{Power Output}}{\text{Power Input}} \leq 1} \tag{14.9}$$

Overall efficiency of machines working in series is equal to the product of their individual efficiencies.

18.5 IMPULSE AND MOMENTUM

Impulse-momentum relationship:

$$\boxed{\begin{gathered}\text{Initial system momenta + Resultant external impulse} \\ = \text{Final system momenta}\end{gathered}} \tag{18.14}$$

Three separate equations may be derived:

A) Equate the x-components of impulse and momenta.

B) Equate the y-components of impulse and momenta.

C) Equate the moments of the vectors about any given point.

For a system of rigid bodies:

A) Consider each body separately.

B) If the problem contains no more than three unknowns, the principle of impulse and momentum may be applied to the system as a whole.

18.6 CONSERVATION OF ANGULAR MOMENTUM

If either

A) the lines of action of all external forces passes through a point O; or

B) the sum of the angular impulses of all external forces are zero,

then:

The angular momentum about a given point is conserved:

$$(\vec{H}_0)_1 = (\vec{H}_0)_2 \tag{18.15}$$

Applications:

A) If rotation of the body is constant (uniform angular velocity)

B) Valid only for the system as a whole, never to its compoment parts.

18.7 ECCENTRIC IMPACT

Unconstrained Case

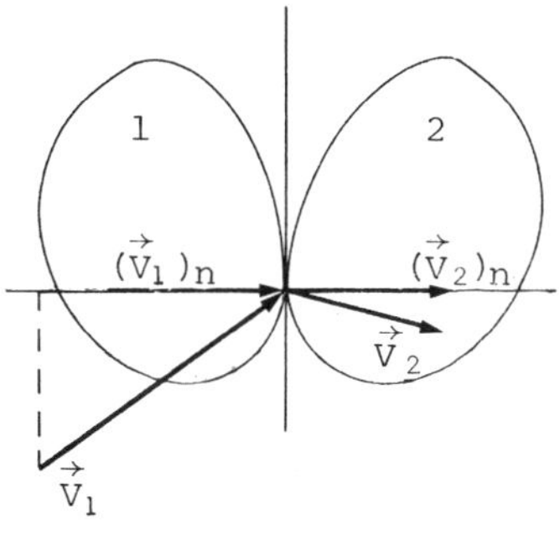

Fig. 18.1

Velocity Relationships:

$$(v_2')_n - (v_1')_n = e[(v_1)_n - (v_2)_n] \qquad (18.16)$$

e = Coefficient of Restitution

Prime (') variables represent parameters after the impact.

Constrained Case

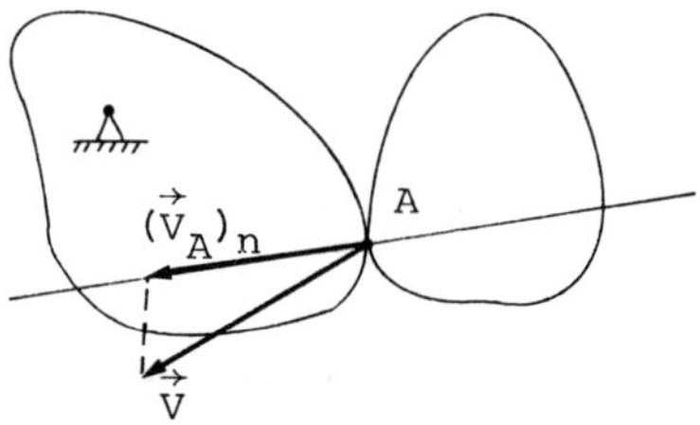

Fig. 18.2

$$e = \frac{(u_A)_n - (v_A')_n}{(v_A)_n - (u_A)_n} \qquad (18.17)$$

where u = Velocity after the period of deformation.

CHAPTER 19

SPATIAL DYNAMICS

19.1 ANGULAR MOMENTUM

Angular momentum of a rigid body about its mass center:

$$\begin{aligned} H_x &= I_{xx}\omega_x - P_{xy}\omega_y - P_{xz}\omega_z \\ H_y &= -P_{yx}\omega_x + I_{yy}\omega_y - P_{yz}\omega_z \\ H_z &= -P_{zx}\omega_x - P_{zy}\omega_y + I_{zz}\omega_z \end{aligned} \tag{19.1}$$

where

$$\begin{aligned} I_{xx} &= \int (y^2+z^2)dm \\ I_{yy} &= \int (z^2+x^2)dm \\ I_{zz} &= \int (x^2+y^2)dm \end{aligned} \tag{19.2}$$

are the centroidal mass moments of inertia; and

$$P_{xy} = \int xydm$$

$$P_{yz} = \int yzdm$$

$$P_{zx} = \int zxdm \qquad (19.3)$$

are the centroidal mass products of inertia. From equation (19.1), we get the inertia tensor at the mass center:

$$\begin{pmatrix} I_{xx} & -P_{xy} & -P_{xz} \\ -P_{yx} & I_{yy} & -P_{yz} \\ -P_{zx} & -P_{zy} & I_{zz} \end{pmatrix} \qquad (19.4)$$

Principal Axes of Inertia - A set of body axes for which the products of inertia are zero.

If the principal axes of inertia were used as reference, then:

Angular Momentum:

$$H_{x'} = I_{xx'}\omega_{x'}$$

$$H_{y'} = I_{yy'}\omega_{y'}$$

$$H_{z'} = I_{zz'}\omega_{z'} \qquad (19.5)$$

$\vec{H}_G$ and $\vec{\omega}_G$ have the same direction if, and only if, $\vec{\omega}$ is directed along a principal axis of inertia.

The inertia tensor corresponding to the principal axes of inertia is defined as:

$$\begin{pmatrix} I_{xx'} & 0 & 0 \\ 0 & I_{yy'} & 0 \\ 0 & 0 & I_{zz'} \end{pmatrix} \qquad (19.6)$$

Transfer Formula:

$$\boxed{\vec{H}_P = \vec{H}_G + \vec{\rho} \times m\vec{v}} \tag{19.7}$$

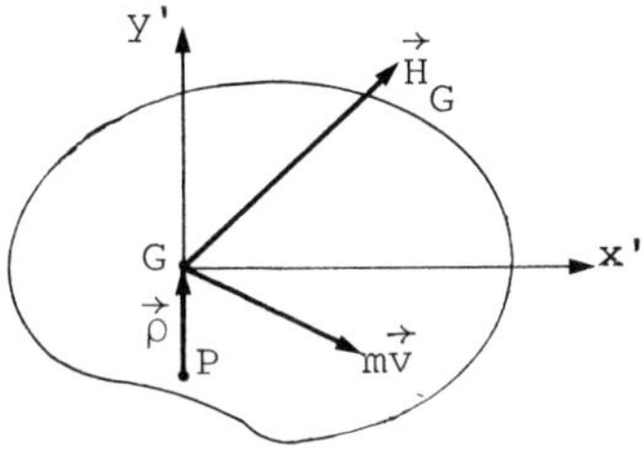

Fig. 19.1

A standard way of writing expression (19.4) is:

$$\boxed{\begin{pmatrix} I_{11} & I_{12} & I_{13} \\ I_{21} & I_{22} & I_{23} \\ I_{31} & I_{32} & I_{33} \end{pmatrix}} \tag{19.8}$$

where

$$I_{xx} = I_{11} \qquad -P_{xy} = I_{12}$$
$$I_{yy} = I_{22} \qquad -P_{xz} = I_{13}$$
$$I_{zz} = I_{33} \qquad -P_{zx} = I_{31}$$
$$\vdots$$
etc.

Note: The moment of inertia terms have repeated subscripts and the product of inertia terms have different subscripts.

The angular momentum expressions, given by (19.1), may be conveniently expressed in matrix notation as:

$$\begin{bmatrix} H_x \\ H_y \\ H_z \end{bmatrix} = \begin{bmatrix} I_{xx} & I_{xy} & I_{xz} \\ I_{yx} & I_{yy} & I_{yz} \\ I_{zx} & I_{zy} & I_{zz} \end{bmatrix} \begin{bmatrix} \omega_x \\ \omega_y \\ \omega_z \end{bmatrix} \tag{19.9}$$

where $I_{xy} = -P_{xy}$, $I_{xz} = -P_{xz}$, $I_{yx} = -P_{yx}$, $I_{yz} = -P_{yz}$, $I_{zx} = -P_{zx}$, and $I_{zy} = -P_{zy}$.

Equation 19.9 may be reduced to the form,

$$\vec{H} = \bar{I}\,\vec{\omega} \tag{19.10}$$

where $\vec{H} = \begin{bmatrix} H_x \\ H_y \\ H_z \end{bmatrix}$ $\quad \bar{I} = \begin{bmatrix} I_{xx} & I_{xy} & I_{xz} \\ I_{yx} & I_{yy} & I_{yz} \\ I_{zx} & I_{zy} & I_{zz} \end{bmatrix}$

and $\vec{\omega} = \begin{bmatrix} \omega_x \\ \omega_y \\ \omega_z \end{bmatrix}$

Note: An alternate form of equation (19.10) may be written using equation (19.8), i.e., in tensor notation as

$$H_\alpha = I_{\alpha\beta}\,\omega_\beta \qquad \alpha,\beta = 1,2,3$$

where the Einstein summation convention is employed, i.e., repeated subscripts in the same term imply summation over the range of that index.

19.2 PRINCIPLE AXES

Principal Axes - A set of coordinate axes in which the products of inertia are zero.

The angular momentum in reference to principal axes is given by equation (19.5) and the inertia tensor is given by equation (19.6).

A device such as a rotating shaft is

A) Statically Balanced - If the center of mass lies on the axis of rotation, and

B) Dynamically Balanced - If the products of inertia are zero and the center of mass lies on the axis of rotation.

Finding the principal axes is equivalent to diagonalizing a (3×3) matrix and is accomplished by solving

$$\left| \bar{I} - \lambda \bar{u} \right| = 0 \tag{19.11}$$

where $\bar{u}$ is a unit matrix and $\bar{I}$ is given by equation (19.10). Equation (19.11) may be expanded as

$$\begin{vmatrix} I_{xx-\lambda} & I_{xy} & I_{xz} \\ I_{yx} & I_{yy-\lambda} & I_{yz} \\ I_{zx} & I_{zy} & I_{zz-\lambda} \end{vmatrix} = 0 \tag{19.12}$$

and reduced to the cubic form as:

$$-\lambda^3 + A\lambda^2 + B\lambda + C = 0 \tag{19.13}$$

where A, B and C are functions of the I's. The roots of equation (19.13) λ_1, λ_2 and λ_3 are the principal moments of inertia.

To find the orientation, the following equation should be solved:

$$\vec{H} = \lambda\vec{\omega} = \bar{I}\,\bar{\omega} \tag{19.14}$$

in which λ is one of the three roots (λ_1, λ_2 or λ_3)

Scalar form of equation (19.14):

$$(I_{xx}-\lambda)\cos\alpha + I_{xy}\cos\beta + I_{xz}\cos\gamma = 0$$

$$I_{yx}\cos\alpha + (I_{yy}-\lambda)\cos\beta + I_{yz}\cos\gamma = 0 \qquad (19.15)$$

$$I_{zx}\cos\alpha + I_{zy}\cos\beta + (I_{zz}-\lambda)\cos\gamma = 0$$

α, β and γ are the direction angles of one of the principal axes. The roots of the equations (19.15) must satisfy:

$$\cos^2\alpha + \cos^2\beta + \cos^2\gamma = 1 \qquad (19.16)$$

If one principal axis is known, the others can be determined by asuming that the z-axis is the principal axis. Then equations (19.15) reduce to

$$(I_{xx}-\lambda)\cos\alpha + I_{xy}\cos\beta = 0$$

$$I_{xy}\cos\alpha + (I_{yy}-\lambda)\cos\beta = 0 \qquad (19.17)$$

Let η be the angle between the x-axis and one of the unknown principal axes, then:

$$\tan 2\eta = \frac{2 I_{xy}}{I_{xx} - I_{yy}} \qquad (19.18)$$

The third axis is normal to this axis: (η +90°).

19.3 MOMENTAL ELLIPSOID

To find the moment of inertia about an arbitrary axis

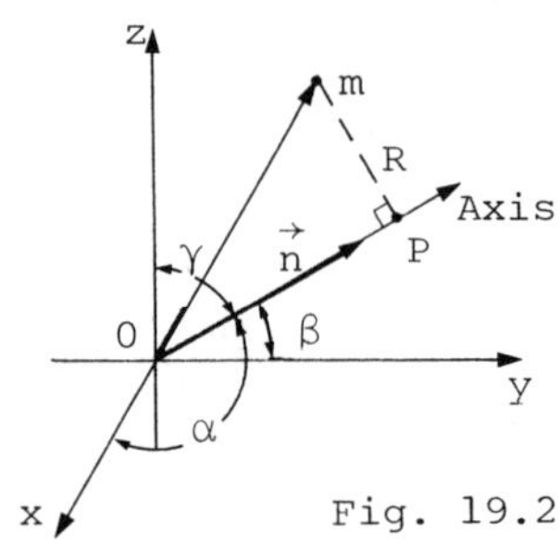

Fig. 19.2

$$I = \Sigma m_j R_j^{\,2}$$ (19.19)

Using direction cosines:

$$I = I_{xx}\cos^2\alpha + I_{yy}\cos^2\beta + I_{zz}\cos^2\gamma + 2I_{yz}\cos\gamma\cos\beta + 2I_{zx}\cos\alpha\cos\gamma + 2I_{xy}\cos\alpha\cos\beta$$ (19.20)

If the coordinate axes are also principal axes at the origin, then equation (19.20) becomes:

$$I = I_{xx}\cos^2\alpha + I_{yy}\cos^2\beta + I_{zz}\cos^2\gamma$$ (19.21)

The abbreviation of equation (19.20) is:

$$I = \bar{n}^T \; \bar{I}\, \bar{n}$$ (19.22)

The distance 'OP' is given by:

$$OP = \frac{1}{\sqrt{I}}$$ (19.23)

By equating the direction cosines in terms of $\sqrt{I}$:

$$\bar{n} = \begin{bmatrix} x\sqrt{I} \\ y\sqrt{I} \\ z\sqrt{I} \end{bmatrix} = \bar{r}\sqrt{I}$$ (19.24)

Substituting into the general formula for the moment of inertia:

$$\bar{r}^T\,\bar{I}\,\bar{r} = 1$$

(19.25)

or

$$x^2 I_{xx} + y^2 I_{yy} + z^2 I_{zz} + 2yzI_{yz} + 2zxI_{zx} + 2xyI_{xy} = 1 \quad (19.26)$$

This is the equation of a surface called the momental ellipsoid of the body at O. If the coordinate axes are the principal axes, then equation (19.26) becomes:

$$x^2 I_{xx} + y^2 I_{yy} + z^2 I_{zz} = 1 \quad (19.27)$$

Note:

A) The principal axes of the body are collinear with the principal axes of the ellipsoid.

B) If two principal moments of inertia are equal, then the ellipsoid is a surface of revolution.

C) If three are equal, then it is a sphere, and the moment of inertia is the same for any axis passing through 'O' regardless of its direction.

19.4 IMPULSE AND MOMENTUM

The relationship as described in section 18.5 is applicable to spatial dynamics also:

$$\text{System Momenta}_1 + \text{System External Impulse}_{1\to 2} = \text{System Momenta}_2 \quad (18.14)$$

Points to Remember

Use figures to write component and moment equations.

Remember that compoments of $\vec{H}_G$ are related to the compoments of $\vec{\omega}$.

For motion about a fixed point in rotation, eliminate reactions at that point by writing an equation involving moments of the momenta and impulses about that point.

19.5 KINETIC ENERGY

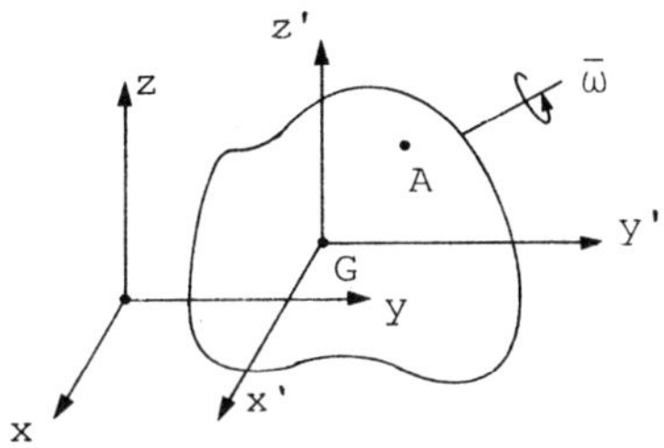

Fig. 19.3

The general expression for kinetic energy is:

$$KE = \frac{1}{2} m V_G^2 + \frac{1}{2} \vec{\omega} \cdot \vec{H}_G + \vec{V}_G \cdot V_{A/G} m \qquad (19.28)$$

If G the mass center is used as the reference point, then

$$KE = \frac{1}{2} m V_{cm}^2 + \frac{1}{2} \vec{\omega} \cdot \vec{H}_G \qquad (19.29)$$

In scalar form (of equation (19.29)):

$$KE = \frac{1}{2} mV_{cm}^2 + \frac{1}{2} (I_x \omega_x^2 + I_y \omega_y^2 + I_z \omega_z^2 - 2P_{xy} \omega_x \omega_y - 2P_{yz} \omega_y \omega_z - 2P_{zx} \omega_z \omega_x) \qquad (19.30)$$

If the axes x'y'z' are also the principal axes, then:

$$T = \frac{1}{2} I_x \omega_x^2 + \frac{1}{2} I_y \omega_y^2 + \frac{1}{2} I_z \omega_z^2 \qquad (19.31)$$

Matrix Notation

$$T = \frac{1}{2} \bar{\omega}^T \bar{I} \bar{\omega}$$

$$= \frac{1}{2} [\omega_x \; \omega_y \; \omega_z] \begin{bmatrix} I_{xx} & I_{xy} & I_{xz} \\ I_{yx} & I_{yy} & I_{yz} \\ I_{zx} & I_{zy} & I_{zz} \end{bmatrix} \begin{bmatrix} \omega_x \\ \omega_y \\ \omega_z \end{bmatrix} \qquad (19.32)$$

19.6 RIGID BODY MOTION

Moment Equation:

$$\Sigma M_A = \vec{r}_{G/A} \times m\vec{a}_A + \dot{\vec{H}}_A \tag{19.33}$$

where A is an arbitrary reference point

$\vec{r}_{G/A}$ is the position vector from A to the mass center G.

General Approach to Solving a Problem:

Reduce a given force system to a resultant force-couple system acting at G. Equation (19.33) becomes:

$$\Sigma \vec{M}_A = \vec{r}_{G/A} \times m\vec{a}_{cm} + \dot{\vec{H}}_G \tag{19.34}$$

To evaluate $\dot{\vec{H}}$:

$$\dot{\vec{H}} = \frac{\delta \vec{H}}{\delta t} + \vec{\omega} \times \vec{H} \tag{19.35}$$

$\dot{\vec{H}}$ = Absolute time rate of change of $\vec{H}$ with respect to the inertial axes

$\frac{\delta \vec{H}}{\delta t} = \dot{\vec{H}}_r$ = Relative time rate of change of $\vec{H}$ with respect to the rotating body axes

$\vec{\omega} \times \vec{H}$ = Rate of change of the direction of $\vec{H}$ due to rotation of the body axes

Component Equations:

$$\begin{aligned} \Sigma M_x &= \dot{H}_x + H_z \omega_y - H_y \omega_z \\ \Sigma M_y &= \dot{H}_y + H_x \omega_z - H_z \omega_x \\ \Sigma M_z &= \dot{H}_z + H_y \omega_x - H_x \omega_y \end{aligned} \tag{19.36}$$

where:

$$H_x = I_x \omega_x - P_{xy} \omega_y - P_{xz} \omega_z$$
$$H_y = I_y \omega_y - P_{yz} \omega_z - P_{yx} \omega_x$$
$$H_z = I_z \omega_z - P_{zx} \omega_x - P_{zy} \omega_y$$

19.7 EULER'S EQUATIONS OF MOTION

By using body axes that coincide with the principal axes of inertia, the moment equation can be simplified to the following known as Euler's equations of motion:

$$\begin{aligned} \Sigma M_x &= I_x \dot{\omega}_x + (I_z - I_y) \omega_y \omega_z \\ \Sigma M_y &= I_y \dot{\omega}_y + (I_x - I_z) \omega_z \omega_x \\ \Sigma M_z &= I_z \dot{\omega}_z + (I_y - I_x) \omega_x \omega_y \end{aligned} \qquad (19.37)$$

The above equations are used to analyze the motion of a rigid body about its mass center.

$$\begin{aligned} \Sigma F_x &= ma_x \\ \Sigma F_y &= ma_y \\ \Sigma F_z &= ma_z \end{aligned} \qquad (19.38)$$

Newton's Laws with Euler's equations completely solve the motion of a rigid body in three dimensions.

19.8 MOTION ABOUT A FIXED POINT

Consider the rate of change of $\vec{H}_0$ about the fixed point O:

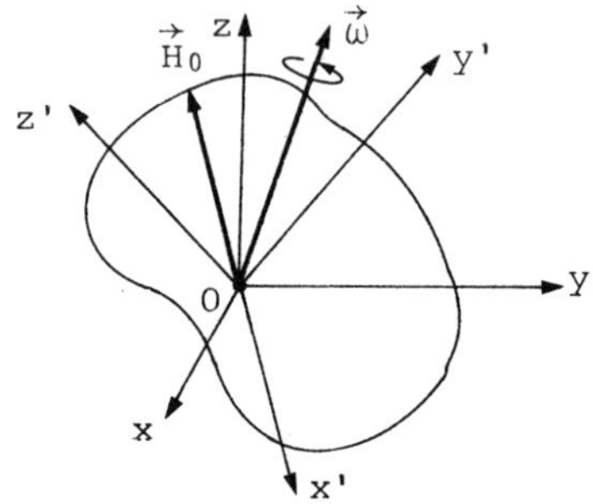

Fig. 19.4

$$\Sigma \vec{M}_0 = \dot{\vec{H}}_0 = (\dot{\vec{H}}_0)_{oxyz} + \vec{\Omega} \times \vec{H}_0 \tag{19.39}$$

$\vec{H}_0$ = Angular momentum with respect to oxyz

$(\dot{\vec{H}}_0)_{oxyz}$ = Rate of change of $\vec{H}_0$ with respect to the rotating frame ox'y'z'

$\vec{\Omega}$ = Angular velocity of ox'y'z'

19.9 MOTION ABOUT A FIXED AXIS

Assume that the z-axis is the axis of rotation:

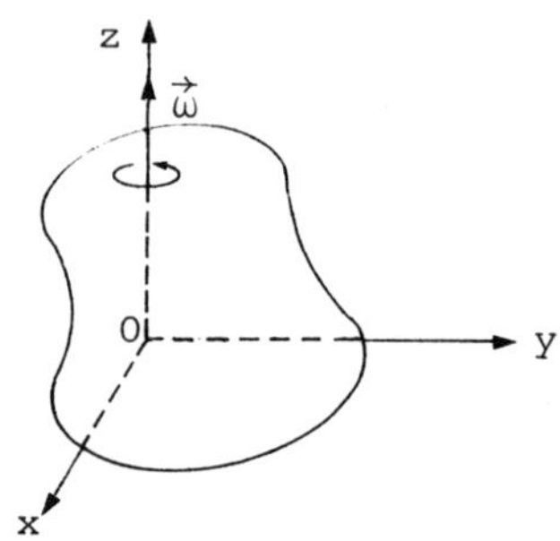

Fig. 19.5

Non-Symmetrical Bodies:

$$\begin{aligned} \Sigma M_x &= -P_{zx}\alpha + P_{yz}\omega^2 \\ \Sigma M_y &= -P_{yz}\alpha - P_{xz}\omega^2 \\ \Sigma M_z &= I_z\alpha \end{aligned} \qquad (19.40)$$

$\alpha = \dot{\omega}_z$ = Angular acceleration in z direction

Symmetrical Bodies:

$$\begin{aligned} \Sigma M_x &= \Sigma M_y = 0 \\ \Sigma M_z &= I_z \dot{\omega}_z = I_z \alpha \end{aligned} \qquad (19.41)$$

Application:

A) When the forces applied are known, use ΣM_z to find α.

B) Determine ω by integration and substitute it into the ΣM_x and ΣM_y equations.

C) Together with the compoment equations of $\Sigma \vec{F} = m\vec{a}$, these equations define the motion of the mass center.

19.10 EARTH'S ROTATIONAL EFFECTS

Static Effects - A study of the plum line. Choose a coordinate system with the origin at the bob.

Equation of motion, equation (16.24) becomes:

$$\boxed{\vec{F} - m\vec{A}_0 = 0} \qquad (19.42)$$

where $\vec{F}$ includes:

1) $m\vec{G}$ = Gravitational Attraction

2) $-m\vec{g}$ = Vertical Tension of the Plum Line and

$\vec{A}_0$ = Centripetal Acceleration of the Moving Origin with a magnitude of $\rho\omega^2$

where $\rho = r\cos\lambda$ (see Fig. 19.6)

The plum line does not point to the earth's center, but deviates by an amount e:

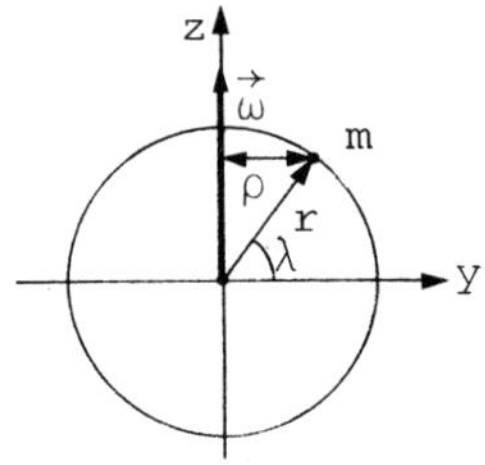

Fig. 19.6

$$\frac{\sin e}{mr\omega^2 \cos\lambda} = \frac{\sin\lambda}{mg} \tag{19.43}$$

Since e is a small quantity, equation (19.43) can be approximated as:

$$\sin e \simeq e = \frac{r\omega^2}{g} \sin\lambda\cos\lambda = \frac{r\omega^2}{2g} \sin 2\lambda \tag{19.44}$$

Maximum deviation occurs at $\lambda = 45°$ where

$$e_{max} = \frac{1}{10} \text{ degree} \tag{19.45}$$

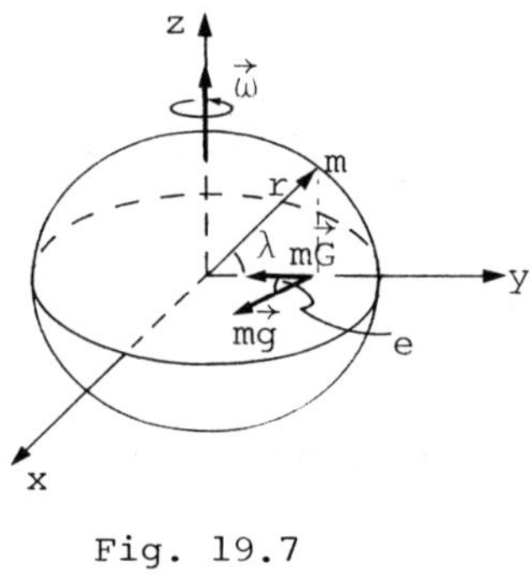

Fig. 19.7

Dynamic Effects - Projectile motion.

Equation of motion:

$$\vec{F} + m\vec{g} - 2m\vec{\omega}\times\dot{\vec{r}} - m\vec{\omega}\times(\vec{\omega}\times\vec{r}) = m\ddot{\vec{r}} \tag{19.46}$$

Neglect air resistance $\vec{F} = 0$ in above equation.

Neglect the term $-m\vec{\omega}\times(\vec{\omega}\times\vec{r})$ as it is very small.

Resulting equation is:

$$m\ddot{\vec{r}} = m\vec{g} - 2m\vec{\omega}\times\dot{\vec{r}} \tag{19.47}$$

Coordinate System:

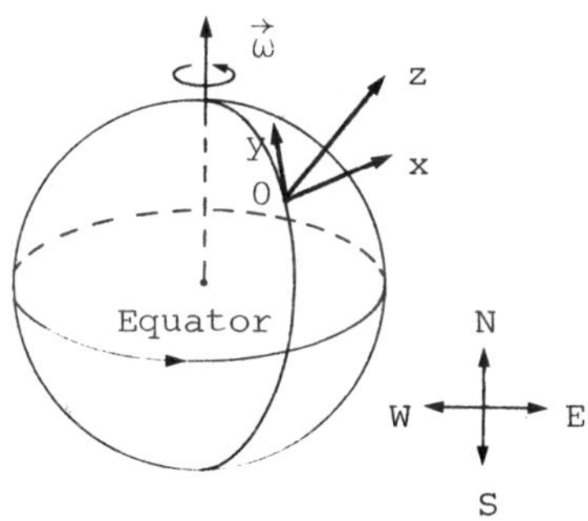

Fig. 19.8

With this system:

$$\vec{g} = -g\vec{k}$$

$$\vec{\omega} = (\omega\cos\lambda)\vec{j} + (\omega\sin\lambda)\vec{k} \tag{19.48}$$

Component Equations of Motion:

$$\ddot{x} = -2\omega(\dot{z}\cos\lambda - \dot{y}\sin\lambda)$$

$$\ddot{y} = -2\omega(\dot{x}\sin\lambda)$$

$$\ddot{z} = -g + 2\omega\dot{x}\cos\lambda \tag{19.49}$$

These equations are solved simultaneously by integration and eliminating ω^2.

$$x = \frac{1}{3}\omega g t^3 \cos\lambda - \omega t^2(\dot{z}_0 \cos\lambda - \dot{y}_0 \sin\lambda) + \dot{x}_0 t$$
$$y = \dot{y}_0 t - \omega \dot{x}_0 t^2 \sin\lambda$$
$$z = -\frac{1}{2} g t^2 + \dot{z}_0 t + \omega \dot{x}_0 t^2 \cos\lambda \qquad (19.50)$$

where $\dot{x}_0, \dot{y}_0, \dot{z}_0$ are the initial conditions.

Special Cases:

A) If the projectile is released from rest in mid air:

Initial Conditions: $\dot{x}_0 = \dot{y}_0 = \dot{z}_0 = 0$

The solution is then

$$x = \frac{1}{3}\omega g t^3 \cos\lambda$$
$$y = 0$$
$$z = -\frac{1}{2} g t^2 \qquad (19.51)$$

The projectile drifts to the east.

B) If the projectile is moving with high velocity in a horizontal direction initially:

$$\dot{x}_0 = v_0, \quad \dot{y}_0 = \dot{z}_0 = 0$$

Equation of Motion:

$$y = -\omega v_0 t^2 \sin\lambda \qquad (19.52)$$

The projectile drifts to the right.

The amount of drift is given by:

$$\text{Drift} = \frac{\omega H^2}{v_0} \sin\lambda \qquad (19.53)$$

where $H \cong v_0 t$, t = Time of Flight

Note: The amount of drift is the same regardless of the initial direction, provided that the trajectory is planar.

19.11 MOTION OF A GYROSCOPE

Consider a body rotating about two axes simultaneously.

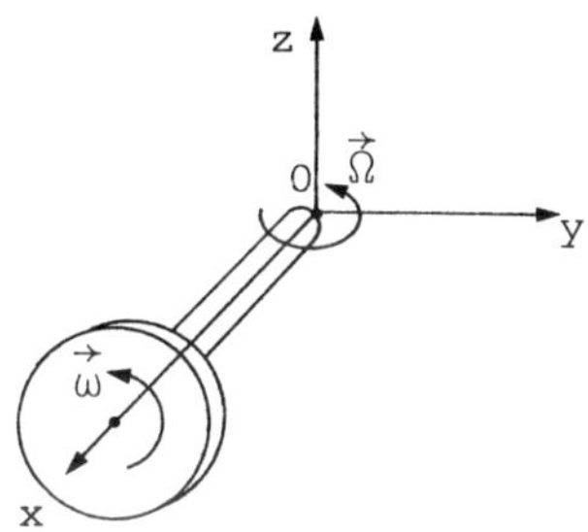

Fig. 19.9 Simple Gyroscope

Nomenclature:

x-Axis = Spin Axis

y-Axis = Torque Axis

z-Axis = Precession Axis

If all the axes are mutually perpendicular to each other, then:

$$\vec{M}_0 = \dot{\vec{H}}_0 \tag{19.54}$$

and

$$M_0 = \vec{I}_{cm}\, \omega \Omega$$

Ω = Rate of Precession

Note: A force applied in the x-y plane trying to increase the precession will result in making the disk and axle rise.

Reference Position of a Gyroscope:

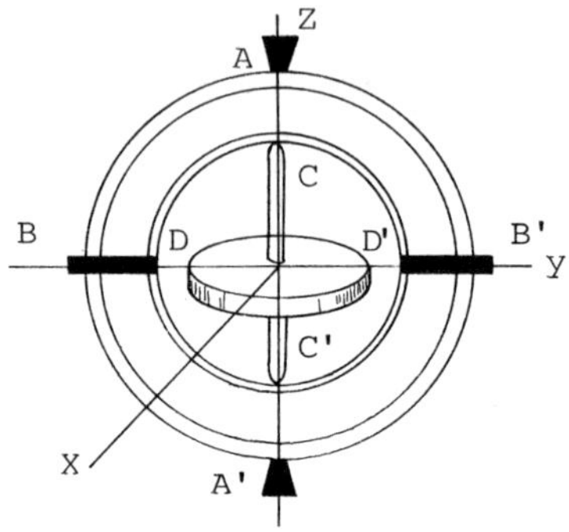

Fig. 19.10

Reference Angles:

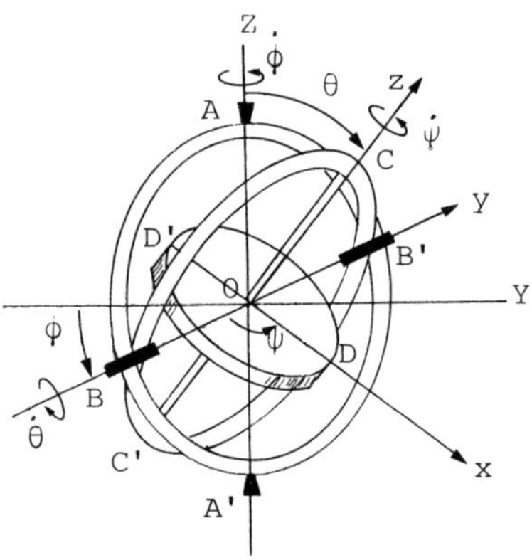

Fig. 19.11

Note:

A) The angles θ, ϕ, ψ and are Eulerian angles. They describe the position of the gyroscope.

B) $\dot{\phi}$ = Rate of Precession

$\dot{\theta}$ = Rate of Nutation

$\dot{\psi}$ = Rate of Spin

C) Angular displacements are not vectors; time derivatives of the angles are vectors.

D) Eulerian angles may be used for any rigid body, in defining its orientation with respect to axes situated at any point of the body.

E) Oxyz are the coordinate axes and are attached to the inner gimbal ring. They are the principal axes of inertia for the gyroscope.

Angular velocity with respect to the fixed axes oxyz:

$$\vec{\omega} = -\dot{\phi}\sin\theta\,\vec{i} + \dot{\theta}\vec{j} + (\dot{\psi} + \dot{\phi}\cos\theta)\vec{k} \tag{19.55}$$

Angular Momentum:

$$\vec{H}_0 = -I'\dot{\phi}\sin\theta\,\vec{i} + I'\dot{\theta}\vec{j} + I(\dot{\psi} + \dot{\phi}\cos\theta)\vec{k} \tag{19.56}$$

where $I = I_z$

$I' = I_x$

Angular velocity with respect to the rotating axes O_{xyz}:

$$\vec{\Omega} = -\dot{\phi}\sin\theta\,\vec{i} + \dot{\theta}\vec{j} + \dot{\phi}\cos\theta\,\vec{k} \tag{19.57}$$

From equation (19.39)

$$\begin{aligned}
\Sigma M_x &= -I'(\ddot{\phi}\sin\theta + 2\dot{\theta}\dot{\phi}\cos\theta) + I\dot{\theta}(\dot{\psi} + \dot{\phi}\cos\theta) \\
\Sigma M_y &= I'(\ddot{\theta} - \dot{\phi}^2\sin\theta\cos\theta) + I\dot{\phi}\sin\theta(\dot{\psi} + \dot{\phi}\cos\theta) \\
\Sigma M_z &= I\frac{d}{dt}(\dot{\psi} + \dot{\phi}\cos\theta)
\end{aligned} \tag{19.58}$$

Solutions of equation (19.58) may be found by numerical methods and digital computers. Analytical solutions have not yet been obtained.

19.12 STEADY PRECESSION

Steady Precession - Special condition in which θ, $\dot{\phi}$, and $\dot{\psi}$ are all constants.

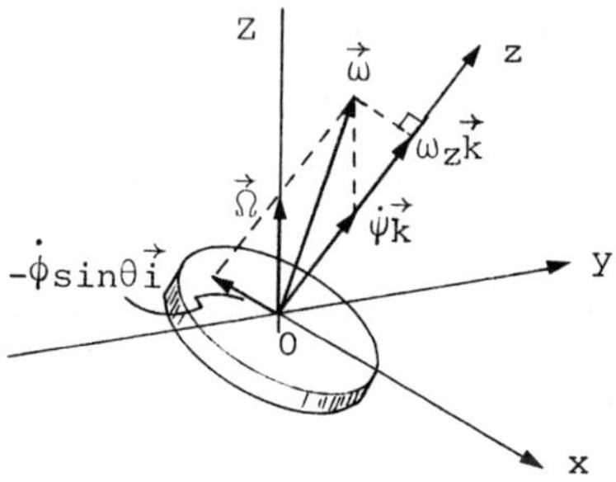

Fig. 19.12

$$\vec{\omega} = -\dot{\phi}\sin\theta\vec{i} + \omega_z\vec{k} \tag{19.59}$$

$$\vec{H}_0 = -I'\dot{\phi}\sin\theta\vec{i} + I\omega_z\vec{k} \tag{19.60}$$

$$\vec{\Omega} = -\dot{\phi}\sin\theta\vec{i} + \dot{\phi}\cos\theta\vec{k} \tag{19.61}$$

and

$$\omega_z = \dot{\psi} + \dot{\phi}\cos\theta \tag{19.62}$$

The couple which must be applied to maintain precession:

$$\boxed{\Sigma\vec{M}_0 = (I\omega_z - I'\dot{\phi}\cos\theta)\dot{\phi}\sin\theta\vec{j}} \tag{19.63}$$

This couple must act about an axis normal to the precession axis and the spin axis of the gyroscope:

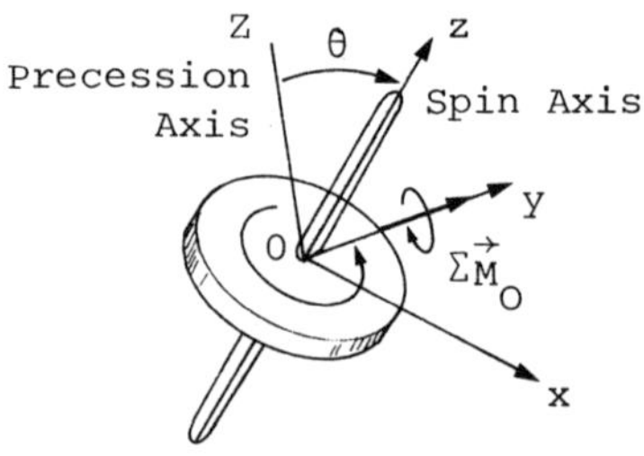

Fig. 19.13

Special Case: When $\theta = 90°$

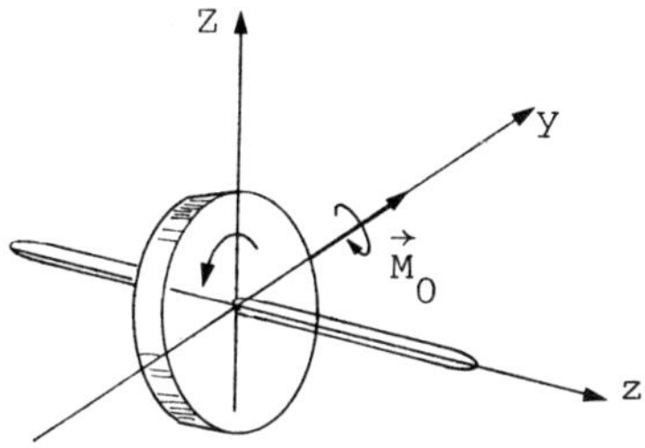

Fig. 19.14

$$\boxed{\Sigma \vec{M}_0 = I \dot{\psi} \dot{\phi} \vec{j}} \tag{19.64}$$

Spinning Top

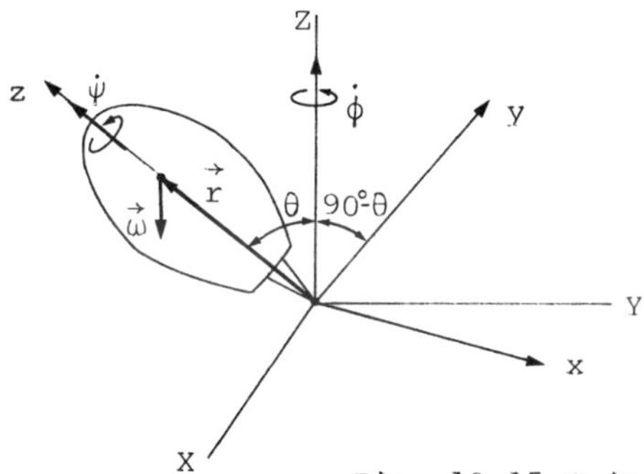

Fig. 19.15 Spinning Top

Equations of Motion:

$$\Sigma M_x = I[(\dot{\psi} + \dot{\phi}\cos\theta)\dot{\phi} - I'\dot{\phi}^2\cos\theta]\sin\theta$$

$$\Sigma M_y = \Sigma M_z = 0 \tag{19.65}$$

The applied moment is due to weight:

$$\Sigma M_x = Wr\sin\theta = [I(\dot{\psi} + \dot{\phi}\cos\theta)\dot{\phi} - I'\dot{\phi}^2\cos\theta]\sin\theta \quad (19.66)$$

Rate of Precession:

$$\dot{\phi} = \frac{I\,\omega_z}{2I'\cos\theta}\left[1 \pm \sqrt{1 - \frac{4WrI\cos\theta}{I^2\omega_z^2}}\right] \qquad (19.67)$$

Two Cases:

A) If $\theta > 90°$:

The top or gyro rotor is called a pendulous gyroscope with ($\cos\theta < 0$) and ϕ is real.

There are two possible values of $\dot{\phi}$:

1) Positive
2) Negative-Retrograde Precession

B) If $\theta < 90°$:

Then $\dot{\phi}$ is real only if

$$1 \geqq \frac{4WrI'\cos\theta}{I^2\,\omega_z^{\,2}}$$

or

$$\omega_z^2 \geqq \frac{4WrI'\cos\theta}{I^2} \qquad (19.68)$$

As ω_z increases, equation (19.65) approaches:

$$\dot{\phi} = \frac{I\omega_z}{2I'\cos\theta}\left[1 \pm \left(1 - \frac{2WrI'\cos\theta}{I^2\,\omega_z^2}\right)\right] \qquad (19.69)$$

The two roots of equation (19.69) approach the limiting values:

1) Rapid Rate

$$\dot{\phi} = \frac{I\,\omega_z}{I'\cos\theta} \qquad (19.70)$$

2) Slow Rate - Occurs usually

$$\dot{\phi} = \frac{Wr}{I\omega_z} \qquad (19.71)$$

Note: To achieve steady precession, the top must be started with $\dot{\theta} = 0$ and $\dot{\phi}$ specified by equation (19.67).

19.13 TORQUE - FREE MOTION

Motion of a spinning body moving with no external moment about its mass center.

i.e. $\Sigma M_G = 0$.

Typical Examples

Orbiting space vehicles, gimbal supported gyroscope with frictionless bearings, etc.

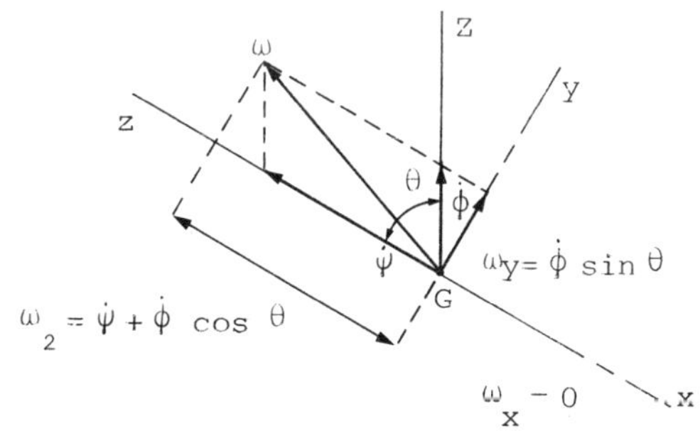

Fig. 19.16

Angular Momentum H_G = Constant as $\Sigma M_G = \dot{H}_G$

$\omega \cdot H_G$ = Constant

Body spin is stabilized if 'ω' and 'H_G' are collinear but does not happen usually (Fig. 19.16).

For steady precession about z-axis

$$\Sigma M_x = 0 \quad \text{and} \quad \dot{\phi} = \frac{I\dot{\psi}}{(I'-I)\cos\theta} \tag{19.72}$$

Relative magnitudes of I and I' determine the direction of precession.

For elongated body where $I' > I$

The positive value of $\dot{\phi}$ produces direct or regular precession.

For a flattened body where $I' < I$

The resulting negative value of $\dot{\phi}$ produces retrograde precession.

CHAPTER 20

LAGRANGIAN MECHANICS

20.1 GENERALIZED COORDINATES AND FORCES

The position of a particle is described by employing the concept of a coordinate system. Given, for example, a coordinate system such as the spherical or the oblate spherical coordinates, etc., a particle in space may be characterized as an ordered triple of numbers called coordinates.

A constrained particle in motion on a surface requires two coordinates, and a constrained particle on a curve, requires one coordinates to characterize its location.

Given a system of 'm' particles, 3M coordinates are required to describe the location of each particle. This is the configuration of the system. (If constraints are imposed on the system, fewer coordinates are required.)

A rigid body requires six coordinates - three for orientation and three for the reference point, to completely locate its position.

Generalized coordinates - A set of coordinates, $q_1, q_2, \ldots, q_m$, equal to the number of degrees of freedom of the system.

If each q_i is independent of the others, then it is known as holonomic.

The rectangular coordinates for a particle expressed in generalized coordinates:

$x = x(q)$ Motion on a curve (one degree of freedom),

$x = x(q_1, q_2)$
$y = y(q_1, q_2)$
motion on a surface (two degrees of freedom),

$x = x(q_1, q_2, q_3)$
$y = y(q_1, q_2, q_3)$
$z = z(q_1, q_2, q_3)$
Spatial motion (three degrees of freedom).

Small Changes in Coordinates:

$$\begin{aligned}
\delta x &= \frac{\partial x}{\partial q_1}\delta q_1 + \frac{\partial x}{\partial q_2}\delta q_2 + \frac{\partial x}{\partial q_3}\delta q_3 \\
\delta y &= \frac{\partial y}{\partial q_1}\delta q_1 + \frac{\partial y}{\partial q_2}\delta q_2 + \frac{\partial y}{\partial q_3}\delta q_3 \qquad (20.1) \\
\delta z &= \frac{\partial z}{\partial q_1}\delta q_1 + \frac{\partial z}{\partial q_2}\delta q_2 + \frac{\partial z}{\partial q_3}\delta q_3
\end{aligned}$$

For a system of 'm' particles in generalized coordinates:

$$\begin{aligned}
\delta x_i &= \frac{\partial x_i}{\partial q_k}\delta q_k \qquad k = 1,2,\ldots,m \quad 1 < i < m \\
\delta y_i &= \frac{\partial y_i}{\partial q_k}\delta q_k \qquad k = 1,2,\ldots,m \quad 1 < i < m \qquad (20.2) \\
\delta z_i &= \frac{\partial z_i}{\partial q_k}\delta q_k \qquad k = 1,2,\ldots,m \quad 1 < i < m
\end{aligned}$$

expressed in tensor notation.

Generalized Forces:

Work done $\delta w = \bar{F} \cdot \delta\bar{r} = F_i \cdot \delta x_i$

For one particle, $1 < i < 3$ and

for 'm' particles, $1 < i < 3m$

In terms of generalized coordinates:

$$\delta w = F_i \frac{\partial x_i}{\partial q_k} \delta q_k \qquad (20.3)$$

or $\delta w = Q_k \delta q_k$, where $Q_k = F_i \frac{\partial x_i}{\partial q_k}$ and is known as the generalized force.

Conservative Systems

Forces expressed in terms of the potential energy function:

$$\boxed{F_i = \frac{-\partial v}{\partial x_i},} \qquad (20.4)$$

where v is the potential energy.

In terms of the generalized force,

$$\boxed{Q_k = -\frac{\partial v}{\partial x_i} \frac{\partial x_i}{\partial q_k} = -\frac{\partial v}{\partial q_k}} \qquad (20.5)$$

20.2 LAGRANGE'S EQUATION

For a system, kinetic energy KE is

$$T = KE = \frac{1}{2} m_i \dot{x}_i^2 \qquad i = 1,2,\ldots,3M \qquad (20.6)$$

where $$\dot{x}_i = \frac{\partial x_i}{\partial q_k} \dot{q}_k + \frac{\partial x_i}{\partial t} \qquad k = 1,2,\ldots,M \qquad (20.7)$$

The Lagrange equation of motion using equations (20.6) and (20.7) is

$$\frac{d}{dt} \frac{\partial T}{\partial \dot{q}_k} = Q_k + \frac{\partial T}{\partial q_k} \qquad k = 1,2,\ldots,M \qquad (20.8)$$

or if the motion is conservative and if the potential energy is a function of generalized coordinates, then equation (20.8) becomes

$$\frac{d}{dt}\left(\frac{\partial T}{\partial \dot{q}_k}\right) = \frac{\partial T}{\partial q_k} - \frac{\partial v}{\partial q_k} \qquad k = 1,2,\ldots,M \qquad (20.9)$$

Lagrange's Function (L)

L = T-V where T and V are in terms of generalized coordinates.

Lagrange's Equation in Terms of L

$$\boxed{\frac{d}{dt}\left(\frac{\partial L}{\partial \dot{q}_k}\right) = \frac{\partial L}{\partial q_k} \qquad k = 1,2,\ldots,M} \qquad (20.10)$$

Lagrange's equation for non-conservative generalized forces:

If $Q_k = Q' - \frac{\partial v}{\partial q_k}$, where Q' is non-conservative then Lagrange's equation becomes

$$\boxed{\frac{d}{dt}\frac{\partial L}{\partial \dot{q}_k} = Q_k' + \frac{\partial L}{\partial q_k},} \qquad (20.11)$$

and is useful, for example, when frictional forces are present.

General Procedure for Obtaining the Equation of Motion:

A) Choose a coordinate system.

B) Write the kinetic energy equation as a function of these coordinates.

C) Find the potential energy, if the system is conservative.

D) Combining these terms in Lagrange's equation results in the equation of motion.

Applications

A) The harmonic oscillator (a non-conservative system), using rectangular coordinates,

Lagrange Function: $L = T - V = \frac{1}{2} m\dot{x}^2 - \frac{1}{2} kx^2$

$$\frac{\partial L}{\partial \dot{x}} = m\dot{x}$$

and

$$\frac{\partial L}{\partial x} = -kx$$

Using the non-conservative version of Lagrange's equation with $Q' = -L\dot{x}$,

$$\frac{d}{dt}(m\dot{x}) = -c\dot{x} + (-kx)$$

and the equation of motion becomes

$$\boxed{m\ddot{x} + c\dot{x} + kx = 0} \qquad (20.12)$$

B) Particle in a Central Field

In terms of polar coordinates: $q_1 = r$; $q_2 = \theta$

$$T = \frac{1}{2} mv^2 = \frac{1}{2} m(\dot{r}^2 + r^2\dot{\theta}^2)$$

$$V = v(r)$$

$$L = \frac{1}{2} m(\dot{r}^2 + r^2\dot{\theta}^2) - v(r)$$

The required partial derivatives are:

$$\frac{\partial L}{\partial \dot{r}} = m\dot{r}; \quad \frac{\partial L}{\partial r} = mr\dot{\theta}^2 - \frac{\partial v}{\partial r} = mr\dot{\theta}^2 + F(r)$$

$$\frac{\partial L}{\partial \theta} = 0 \quad \text{and} \quad \frac{\partial L}{\partial \dot{\theta}} = mr^2\dot{\theta}$$

where $F(r)$ is a function of r equal to $-\partial v/\partial r$. Substituting in Lagrange's equation, the equations of motion become

$$\boxed{\begin{aligned} m\ddot{r} &= mr\dot{\theta}^2 + F(r) \\ \frac{d}{dt}(mr^2\dot{\theta}) &= 0 \end{aligned}} \qquad (20.13)$$

20.3 GENERALIZED MOMENTUM

Momentum P in terms of kinetic energy T:

$$P = \frac{\partial T}{\partial \dot{x}} = m\dot{x} \tag{20.14}$$

Generalized momentun is defined as the momentum P in terms of the generalized coordinates:

$$P_k = \frac{\partial L}{\partial \dot{q}_k} \tag{20.15}$$

For a conservative system equation (20.15) becomes:

$$\dot{P}_k = \frac{\partial L}{\partial q_k} \tag{20.16}$$

20.4 IMPULSIVE FORCES

Given a system with all generalized forces being non-zero for a very small time interval Δt, then the generalized impulse equation is

$$\operatorname*{Lt}_{\Delta t \to o} \int_o^{\Delta t} Q_k dt = \hat{P}_k$$

and

$$\int_o^{\Delta t} \left(\frac{\partial T}{\partial q_k} \right) dt \text{ approaches zero.}$$

The impulse equation may be written as

$$\Delta \left[\frac{\partial T}{\partial \dot{q}_k} \right] = \hat{P}_k \tag{20.17}$$

If the potential energy does not involve q's explicitly,

$$\Delta P_k = \hat{P}_k \tag{20.18}$$

$\hat{P}_K$ is calculated from the impulsive work ($\delta \hat{w}$) equation

$$\begin{aligned} \delta \hat{w} &= \hat{P} \ \delta q_1 + \hat{P}_2 \ \delta q_2 + \ldots \\ &= \hat{P}_k \ \delta q_k \end{aligned} \tag{20.19}$$

20.5 HAMILTON'S VARIATION PRINCIPLE

Hamilton's Variation Principle - The motion of any system occurs such that

$$\int_{t_a}^{t_b} L dt \tag{20.20}$$

takes an extremal value.

More precisely,

$$\delta \int_{t_a}^{t_b} L dt = 0 \tag{20.21}$$

where δ = a variation due to different paths and

L = the Lagrangian function.

20.6 HAMILTON'S EQUATIONS

The Hamiltonian function is defined as

$$H = P_k \dot{q}_k - L \tag{20.22}$$

H will always be a function of P_k, q_k and t. From equation (20.22) we get

$$dH = P_k d\dot{q}_k + \dot{q}_k dP_k - \frac{\partial L}{\partial q_k} dq_k - \frac{\partial L}{\partial \dot{q}_k} d\dot{q}_k - \frac{\partial L}{\partial t} dt$$

Since $P_k = \dfrac{\partial L}{\partial \dot{q}_k}$

$$dH = \dot{q}_k dP_k - \frac{\partial L}{\partial q_k} dq_k - \frac{\partial L}{\partial t} dt \tag{20.23}$$

Because $H = H(P_k, q_k, t)$,

$$dH = \frac{\partial H}{\partial P_k} dP_k + \frac{\partial H}{\partial q_k} dq_k + \frac{\partial H}{\partial t} dt \tag{20.24}$$

Equating (20.24) and (20.23) and since $\dot{P}_k = \dfrac{\partial L}{\partial q_k}$

$$\dot{q}_k = \frac{\partial H}{\partial P_k}; \qquad \dot{P}_k = -\frac{\partial H}{\partial q_k} \tag{20.25}$$

Equations (20.25) are known as Hamilton's Canonical Equations of motion.

20.7 LAGRANGE'S EQUATIONS WITH CONSTRAINTS

Holonomic Constraint - Constraints of the form

$$\boxed{\frac{\partial q}{\partial q_k}\,\delta q = 0} \tag{20.26}$$

Non-holonomic Constraint - Constraints of the form

$$\boxed{h_k\,\delta q_k = 0} \tag{20.27}$$

Differential equations of motion by the method of undetermined multipliers:

(The Non-Holonomic Case)

Multiply the equation by a constant λ and add the result to the integrand of

$$\int_{t_a}^{t_b} \left[\frac{\partial L}{\partial q_k} - \frac{d}{dt}\frac{\partial L}{\partial \dot{q}_k}\right] \delta q_k dt = 0$$

Select λ such that the term in brackets equals zero,

$$\boxed{\begin{gathered} \frac{\partial L}{\partial q_k} - \frac{d}{dt}\frac{\partial L}{\partial \dot{q}_k} + \lambda h_k = 0 \quad (k=1,2,\ldots,m) \\ h_k \dot{q}_k = 0 \end{gathered}} \tag{20.28}$$

There now exist m+1 equations to obtain m+1 unknowns, i.e., $q_1, q_2, \ldots, q_n$, λ.

This technique may be employed with moving constraints or with several constraints by having corresponding undetermined coefficients with corresponding h's in the Lagrangian equations.

CHAPTER 21

MECHANICAL VIBRATIONS

21.1 INTRODUCTION

Nomenclature and Definitions

Mechanical Vibration - The oscillation of a body about an equilibrium position.

Period - Time taken by the system to complete one cycle.

Frequency - The number of cyles per unit time.

Amplitude - Maximum displacement from the equilibrium position.

Free Vibration - Motion is sustained by restoring forces only.

Forced Vibration - Caused by periodic forces applied external to the system.

Undamped Vibration - Friction forces are neglected.

Damped Vibration - Internal and external frictional forces and other damping forces are included.

21.2 SIMPLE HARMONIC MOTION

Simple Harmonic Motion - Linear motion of a body where the acceleration is proportional to the displacement from a fixed origin and is always directed towards the origin. The direction of acceleration is always opposite to that of the displacement.

Equation of Motion

$$m\ddot{x} + kx = 0$$

or

$$\ddot{x} + p^2x = 0 \tag{21.1}$$

where $p^2 = k/m$.

General Solution of Equation (21.1)

$$x = c_1 \sin pt + c_2 \cos pt \tag{21.2}$$

where c_1 and c_2 may be obtained from initial conditions.

An alternate form of equation (21.2)

$$x = x_m \sin(pt + \phi) \tag{21.3}$$

where x_m = the amplitude, and ϕ is the phase angle.

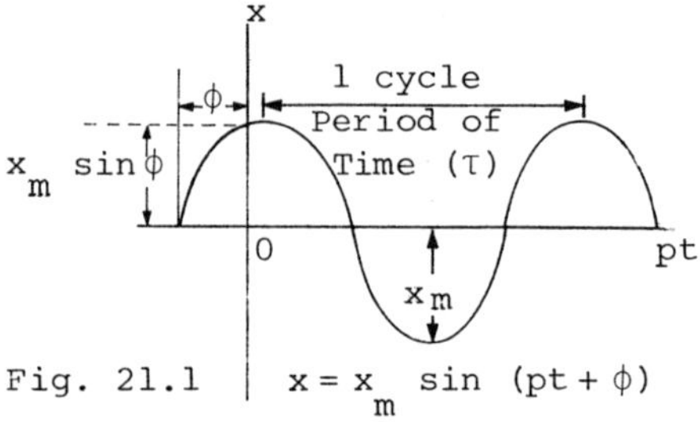

Fig. 21.1

$$\text{Period} = \tau = 2\pi/P$$

$$\text{Frequency, } f = 1/\tau = P/2\pi$$

Maximum Values:

$$V_m = X_m p \qquad a_m = X_m p^2 \tag{21.4}$$

21.3 GENERAL HARMONIC OSCILLATOR

Two-Dimensional Oscillator

Component equation of motion:

$$m\ddot{x} = -kx \qquad m\ddot{y} = -ky \tag{21.5}$$

Solution of equation (21.5) yields

$$x = c_1\cos(\omega t + \theta_1) \qquad y = c_2\cos(\omega t + \theta_2) \qquad \omega = \sqrt{\frac{k}{m}} \tag{21.6}$$

The constants c_1, c_2, θ_1 and θ_2 are determined from initial conditions.

Elimination of 't' between x and y yields the path equation:

$$\frac{x^2}{c_1^2} - xy\frac{2\cos\phi}{c_1c_2} + \frac{y^2}{c_2^2} = \sin^2\phi \quad \text{where } \phi = \theta_2 - \theta_1$$

Because the discriminant is less that zero, the path is an ellipse.

If $\phi = \pi/2$, then the path is an ellipse whose axes coincide with the coordinate axes and the equation is given by

$$\frac{x^2}{c_1^2} + \frac{y^2}{c_2^2} = 1$$

If $\phi = 0$, $y = B/A\ X$

If $\phi = \pi$, $y = -B/A\ X$

Than angle of inclination ψ can be obtained from the relation:

$$\boxed{\tan 2\psi = \frac{2c_1c_2 \cos\phi}{c_1^2 - c_2^2}} \qquad (21.7)$$

Three–Dimensional Oscillator

Equations of Motion are:

$$\boxed{\begin{aligned} m\ddot{x} &= -kx \\ m\ddot{y} &= -ky \\ m\ddot{z} &= -kz \end{aligned}} \qquad (21.8)$$

with corresponding solutions as below:

$$\boxed{\begin{aligned} x &= C_1 \sin\omega t + D_1 \cos\omega t \\ y &= C_2 \sin\omega t + D_2 \cos\omega t \\ z &= C_3 \sin\omega t + D_3 \cos\omega t \end{aligned}} \qquad (21.9)$$

or in vector notation,

$$\vec{r} = \vec{C} \sin\omega t + \vec{D} \cos\omega t$$

Motion is in the plane of the vectors $\vec{C}$ and $\vec{D}$ and the path is an ellipse.

Nonisotropic Oscillator

The restoring force is a function of the direction of displacement.

Equations of Motion:

$$\boxed{\begin{aligned} m\ddot{x} &= -k_a x \\ m\ddot{y} &= -k_b y \\ m\ddot{z} &= -k_c z \end{aligned}} \qquad (21.10)$$

With three corresponding frequencies being

$$\omega_a = \sqrt{k_a/m}$$

$$\omega_b = \sqrt{k_b/m}$$

$$\omega_c = \sqrt{k_c/m}$$

and three corresponding solutions

$$\boxed{\begin{aligned} x &= c_1\cos(\omega_a t + \theta_1) \\ y &= c_2\cos(\omega_b t + \theta_2) \\ z &= c_3\cos(\omega_c t + \theta_3) \end{aligned}} \qquad (21.11)$$

The motion is confined to a cuboid with sides $2c_1$, $2c_2$, and $2c_3$ situated at the origin.

21.4 SIMPLE PENDULUM

The simple pendulum consists of a weightless cord fixed at one end and has a particle at the other end that oscillates in an arc of a circle. The forces involved are gravity and tension. (See Fig. 21.2)

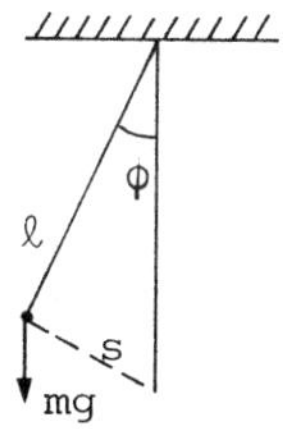

Fig. 21.2

For small angles of vibration

$$\phi = s/\ell$$

Equation of Motion in Terms of ϕ:

$$\ddot{\phi} + \frac{g}{\ell}\sin\phi = 0 \qquad (21.12)$$

If ϕ (in radians) is small, then $\sin\phi \approx \phi$. Equation (21.12) then becomes

$$\ddot{\phi} + \frac{g}{\ell}\phi = 0 \qquad (12.13)$$

The solution for equation (21.13) is given by

$$\phi = \phi_0 \cos(\omega_0 t + \alpha_0)$$

where $\omega_0 = \sqrt{g/\ell}$; ϕ_0 = Max. amplitude of oscillation and α_0 = Phase factor.

The period of oscillation is given by

$$\tau_0 = \frac{2\pi}{\omega_0} = 2\pi\sqrt{\frac{\ell}{g}}$$

Exact Solution

$$\dot{\alpha}^2 = \frac{g}{\ell}\left(1 - \sin^2\frac{\alpha_0}{2}\sin^2\alpha\right) \qquad (21.14)$$

Let $K = \sin^2\frac{\alpha_0}{2}$

$$\dot{\alpha}^2 = \frac{g}{\ell}(1 - K^2\sin^2\alpha)$$

$$\dot{\alpha} = \sqrt{\frac{g}{\ell}}\sqrt{(1 - K^2\sin^2\alpha)} \qquad (21.15)$$

Solving for τ

$$\tau = \sqrt{\frac{\ell}{g}}\int_0^{\phi}\frac{d\phi}{\sqrt{1-K^2\sin^2\alpha}} = \sqrt{\frac{\ell}{g}}\,F(K,\alpha) \qquad (21.16)$$

where $F(k,\alpha)$ is known as the incomplete elliptic integral of the first kind.

Period

$$\tau = 4\sqrt{\frac{\ell}{g}} \int_0^{\pi/2} \frac{d\phi}{\sqrt{1-K^2\sin^2\alpha}} = 4\sqrt{\frac{\ell}{g}}\; K(k) \qquad (21.17)$$

where $K(k) = F(K,\pi/2)$ known as the complete elliptic integral of the first kind.

Numerical values of elliptic integrals can be obtained from tables.

21.5 THE SPHERICAL PENDULUM

The spherical pendulum may oscillate in space as shown in Fig. 21.3.

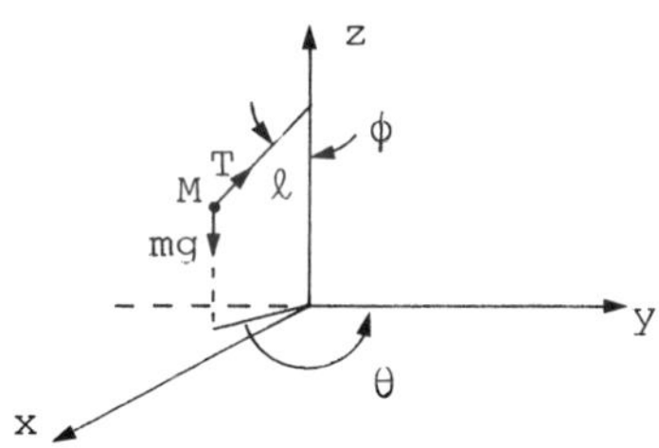

Fig. 21.3

Approximate solution in cartesian coordinates:

Equation of motion:

$$M\ddot{\vec{r}} = M\vec{g} + \vec{T} \qquad (21.18)$$

or in compoment form:

$$\begin{aligned} m\ddot{x} &= T_x \\ m\ddot{y} &= T_y \\ m\ddot{z} &= T_z - mg \end{aligned} \qquad (21.19)$$

Assuming that the displacement from equilibrium is very small:

$$T_x \approx -Mg\, x/\ell$$

$$T_y \approx -Mg\, y/\ell$$

The equations of motion become:

$$\boxed{\begin{aligned} \ddot{x} + (g/\ell)x &= 0 \\ \ddot{y} + (g/\ell)y &= 0 \end{aligned}} \qquad (21.20)$$

with solutions:

$$\boxed{\begin{aligned} x &= A\cos(\omega t + r) \\ y &= B\cos(\omega t + \delta) \end{aligned}} \qquad (21.21)$$

where $\omega = \sqrt{g/\ell}$

On the x-y plane, the motion is an ellipse.

Spherical Coordinates - More accurate than the previous solution.

Equations of Motion:

$$\boxed{\begin{aligned} ma_r &= F_r = Mg\cos\theta - T \\ ma_\phi &= F_\phi = -Mg\sin\phi \\ ma_{\phi\theta} &= F_{\phi\theta} = 0 \end{aligned}} \qquad (21.22)$$

In spherical coordinates:

$$\boxed{\begin{aligned} \ddot{\phi} - \dot{\psi}^2 \sin\phi\cos\phi + g\sin\phi &= 0 \\ \frac{1}{\sin\phi}\frac{d}{dt}(\dot{\alpha}\sin^2\phi) &= 0 \end{aligned}} \qquad (21.23)$$

The $(\dot{\alpha}\sin^2\phi)$ term is constant and is known as the angular momentum 'h':

$$\dot{\alpha} = h/\sin^2\phi$$

Equation in ϕ direction:

$$\boxed{\ddot{\phi} + g/\ell \sin\phi - h^2 \frac{\cos\phi}{\sin^2\phi} = 0} \qquad (21.24)$$

Special Cases:

A) α = Constant:

Equation of motion becomes:

$$\boxed{\ddot{\phi} + g/\ell \sin\phi = 0} \qquad (21.25)$$

which is the case of the simple pendulum discussed in the previous section.

B) $\phi = \phi_0$ = Constant: known as the conical pendulum

Equation of motion:

$$\boxed{h^2 = g/\ell \sin^4\phi_0 \sec\phi_0} \qquad (21.26)$$

Condition for Conical Motion of the Pendulum:

$$\dot{\alpha}_0^2 = \frac{g}{\ell}\sec\phi_0$$

For nearly conical motion ϕ is close to ϕ_0.

We define μ as $\mu = \phi - \phi_0$

Expanding in series and neglecting higher order terms,

$$\ddot{\mu} + \frac{g}{\ell}(\sec\phi_0 + 3\cos\phi)\mu = 0$$

The resulting motion is then defined by the following equation:

$$\boxed{\mu = \phi - \phi_0 = \mu_0\cos\left(\sqrt{\frac{2A}{\ell}}\,t + t\right)} \qquad (21.27)$$

with period:

$$T = 2\pi\sqrt{\frac{\ell}{gA}} \tag{21.28}$$

where $A = \sec\phi_0 + 3\cos\phi_0$.

Energy Considerations

$$V = -Mg\ell\cos\phi$$
$$T = \frac{1}{2}Mv^2 = \frac{1}{2}M(\ell^2\dot{\phi}^2 + \ell\dot{\alpha}^2\sin^2\phi) \tag{21.29}$$

Energy Equation:

$$E = \frac{M\ell^2}{2}(\dot{\phi}^2 + \dot{\alpha}^2\sin^2\phi) - Mg\ell\cos\phi \tag{21.30}$$

Solving for $\dot{\phi}^2$:

$$\dot{\phi}^2 = \frac{2E}{M\ell^2} + \frac{2g}{\ell}\cos\phi - \frac{h^2}{1-\cos^2\phi} = f(\cos\phi) \tag{21.31}$$

The values for which $f(\cos\phi) = 0$ sets the limits of oscillation in ϕ, and is limited for the values of ϕ such that $f(\cos\phi) > 0$. The vertical oscillation is confined therefore between two horizontal circles.

21.6 THE TORSION PENDULUM

The torsion pendulum consists of an elastic rod or wire clamped at one end and having a body attached to its other end such that the axis of the rod passes through the center of gravity of the body (Fig. 12.4).

External moment applied:

$$\boxed{M = \frac{JG}{\ell}\phi} \tag{21.32}$$

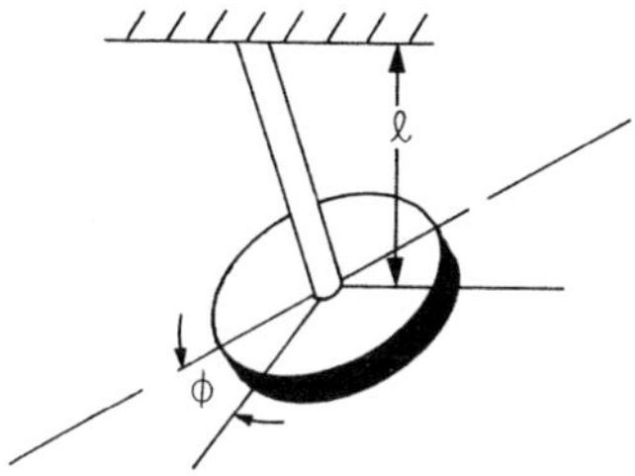

Fig. 21.4

where J = polar moment of inertia; for a rod of diameter α, $J = \pi d^4/32$; G = shear modulus of elasticity; ℓ = length of the rod; and JG/ℓ = Torsional spring constant.

Equation of motion:

$$\boxed{\alpha = -\frac{JG}{I\ell}\phi} \tag{21.33}$$

and with period:

$$\tau = 2\pi\sqrt{\frac{I\ell}{JG}} \tag{21.34}$$

and I is the mass moment of inertia of the body about the axis of rotation.

21.7 THE PHYSICAL PENDULUM

The physical or compound pendulum consists of a rigid body that rotates freely about a fixed horizontal axis and also swings under gravitational influence. Point Q in Fig. 21.5 is known as the center of oscillation, Point C, the center of suspension, and G is the center of gravity.

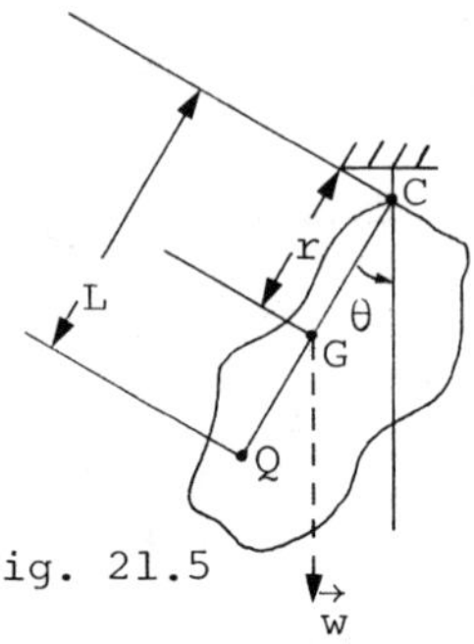

Fig. 21.5

Equation of Motion:

Applying $\Sigma M_x = I_x \alpha$,

or

$$-wr\sin\theta = I_x \alpha$$

$$-wr\sin\theta = \frac{w}{g} K_x^2 \alpha \qquad (21.35)$$

where k_x is the radius of gyration.

If θ is small, the $\sin\theta \approx \theta$, and

$$\alpha = -\frac{gr}{K_x^2}\theta \qquad (21.36)$$

period τ is given by

$$\tau = 2\pi\sqrt{\frac{K_x^2}{gr}} = 2\pi\sqrt{\frac{L}{g}} \qquad (21.37)$$

Note: The period given here equals that of a simple pendulum having a length L equal to K_x^2/r.

With the transfer relation

$$K_x^2 = \overline{K}^2 + r^2$$

the length L may now be expressed as

$$L = \frac{\overline{K}^2}{r} + r \tag{21.38}$$

Therefore, this result implies that for any parallel axis of suspension situated at the same distance from the center of gravity, the period remains the same.

If a body can be made to swing as a pendulum, then equation (21.37) may be employed to experimentally determine its radius of gyration or moment of inertia.

21.8 FREE VIBRATION

Consider the spring-mass system in Fig. 21.6 in which the mass, after being displaced, vibrates freely and indefinitely, neglecting all resistances. Such a condition is known as free vibration.

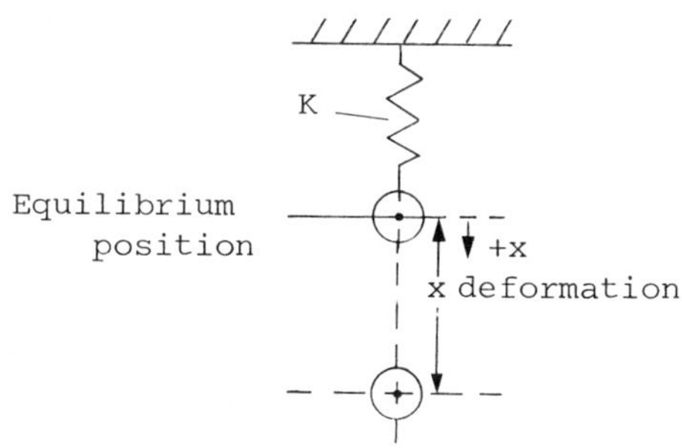

Fig. 21.6

Using Newton's second law, the equation of motion becomes

$$\frac{d^2x}{dt^2} = -\omega^2 x, \tag{21.39}$$

where $$\omega = \sqrt{\frac{Kg}{W}}$$

General solution of equation (21.39)

$$x = A \sin \omega t + B \cos \omega t \tag{21.40}$$

Constants A and B are determined from initial conditions.

Period

$$\tau = 2\pi \sqrt{\frac{W}{Kg}} \tag{21.41}$$

W = Weight of Body

Note: τ depends only on the weight of the body and the spring constant K.

Frequency:

$$f = \frac{1}{2\pi} \sqrt{\frac{Kg}{W}} \tag{21.42}$$

Specifying initial conditions as x_0 and v_0 at $t = 0$, equation (21.40) becomes,

$$\begin{aligned} x &= \frac{v_0}{\omega} \sin \omega t + x_0 \cos \omega t \\ &= C \cos(\omega t - \alpha), \end{aligned} \tag{21.43}$$

where $C = \sqrt{x_0^2 + \left(\frac{v_0}{\omega}\right)^2}$ and α is the phase angle,

equal to $\tan^{-1}(v_0/\omega x_0)$.

Equation (21.43) may be graphically interpreted as in Fig. (21.7).

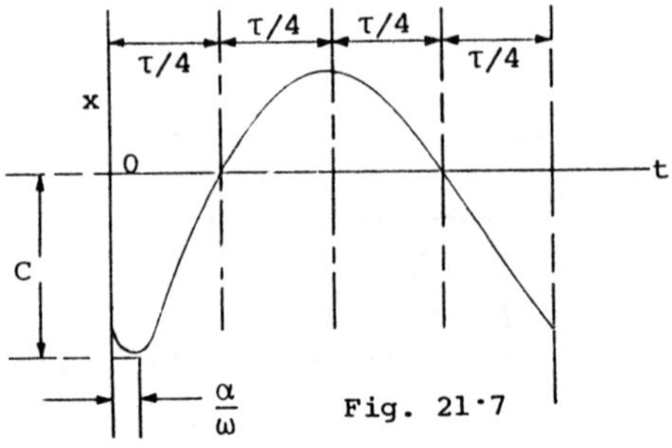

Fig. 21·7

21.9 CONSERVATION OF ENERGY

The method of conservation of energy involves the determination of two quantities:

A) The kinetic energy at the static equilibrium position, and

B) The stored elastic energy at the moment of maximum displacement.

These two quantities are then equated.

For the spring-mass system discussed in section (21.8) the work-energy relation becomes:

$$\boxed{\frac{1}{2}Kx_{max}^2 = \frac{W}{2g}(v^2 + v_{max}^2)} \qquad (21.44)$$

and is reduced further since v = 0 at the maximum displacement,

$$\boxed{\frac{1}{2}Kx_{max}^2 = \frac{W}{2g}(v_{max}^2)} \qquad (21.45)$$

$x = x_m \sin\omega t$

and $v = \omega x_m \cos\omega t$

which gives $v_m = \omega x_m$

From the above relations and substituting them in equation (21.45) we get

$$\omega^2 = \frac{Kg}{W} \qquad (21.46)$$

21.10 FORCED VIBRATION

Forced vibrations occur when an external disturbing force acts on the body continuously and periodically (see Fig. 21.8).

Equation of motion (from Newton's second law):

$$\frac{d^2x}{dt^2} = -\omega^2 x + \frac{F_0}{m}\cos\omega_f t \qquad (21.47)$$

where ω_F = Forcing frequency, F_0 which is the maximum magnitude of the external force, and x is the distance taken from the equilibrium position.

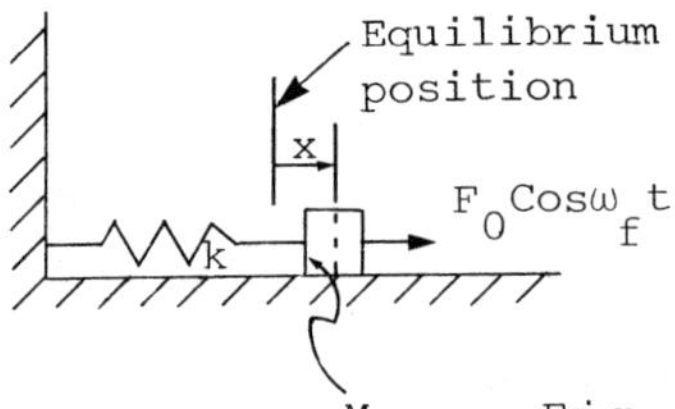

Fig. 21.8

The solution of equation (21.47) consists of the sum of two parts:

A) The complementary solution, X_c and

B) The particular solution, X_p:

The complementary solution is obtained by setting the $\frac{F_0}{m}\cos\omega_f t$ term to zero and

$$x_c = A\cos\omega t + B\sin\omega t \qquad (21.48)$$

The particular solution is found by assuming that x_p = $D\cos\omega_f t$ and substituting it in equation (21.47) to obtain the unkown constant D:

$$D = \frac{F_0}{m} \cdot \frac{1}{\omega^2 - \omega_f^2}$$

Therefore, the solution of equation (21.47) is

$$X = A\cos\omega t + B\sin\omega t + \frac{F_0/m}{\omega^2 - \omega_f^2}\cos\omega_f t \qquad (21.49)$$

Note: The forcing function may be any combination of periodic functions, not neccessarily as simple as $F_0 \cos \omega_f t$.

The magnification factor is defined as the ratio of the amplitude of the steady state vibration D to the static deflection F_0/K (see Fig. 21.9),

$$\text{Magnification Factor} = \frac{1}{1 - \left(\frac{\omega_f}{\omega}\right)^2} \qquad (21.50)$$

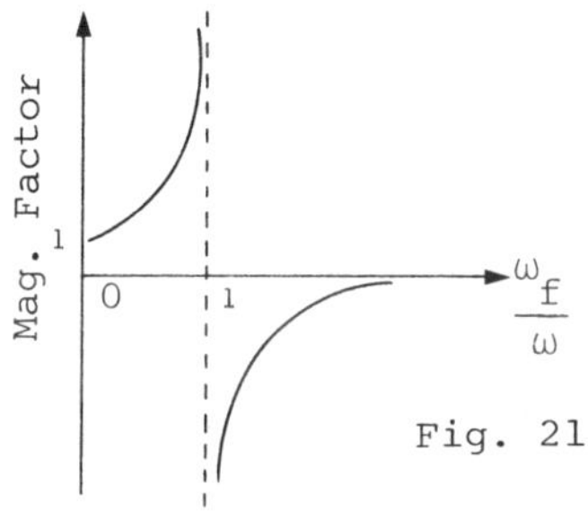

Fig. 21.9

When $\omega_f \approx \omega$, the condition is called resonance and the amplitude of vibration is very large causing quick failure of structures such as bridges or mechanical elements.

When $\omega_f > \omega$, the magnification factor is negative and the disturbing force is in opposite direction to the displacement.

When ω_f is significantly larger than ω, the magnification factor is close to zero and the body is nearly stationary.

21.11 DAMPED FREE VIBRATION

In general, damping is caused by

A) Internal friction,

B) Dry friction between two surfaces,

C) Viscous damping.

The third is the most common in vibrating systems, and is caused by fluid friction. It may be expressed as

$$\vec{F} = c \frac{dx}{dt} \quad (21.51)$$

It is assumed that the body is moving at low velocities (the constant c is the coefficient of viscous damping). An example is the viscous damping provided by the dashpot device as shown in Fig. 21.10.

The equation of motion is:

$$M \frac{d^2x}{dt^2} + c \frac{dx}{dt} + kx = 0 \quad (21.52)$$

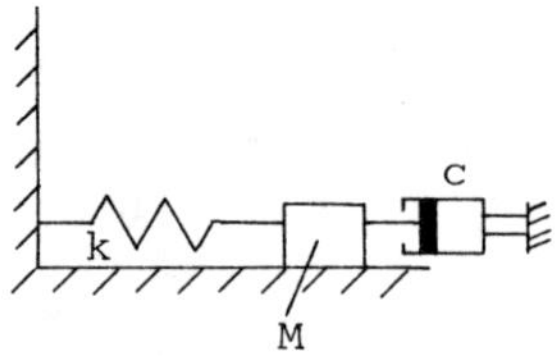

Spring – mass system with a dashpot damper

Fig. 21.10

Equation (21.52) has exponential solutions of the form $x = e^{at}$. Substituting this in equation (21.52) yields two values for 'a' given by:

$$a_1 = \frac{-c}{2m} - \sqrt{\left(\frac{c}{2m}\right)^2 - \frac{K}{M}}$$

$$a_2 = \frac{-c}{2m} + \sqrt{\left(\frac{c}{2m}\right)^2 - \frac{K}{M}}$$

The critical damping coefficient is defined as

$$C_{critical} = 2M\sqrt{\frac{K}{M}} \tag{21.53}$$

Now, three cases must be considered with respect to $C_{critical}$:

A) If $c > C_{critical}$, a_1 and a_2 are both real, the motion is nonoscillating, and the system is overdamped. The general solution is given by

$$x = Ae^{a_1 t} + Be^{a_2 t} \tag{21.54}$$

B) If $c = C_{critical}$, $a_1 = a_2$ and the system is critically damped with general solution:

$$x = (A+Bt)e^{-\left(\frac{C_{critical}}{2m}\right)t} \tag{21.55}$$

C) If $c < C_{critical}$ a_1 and a_2 are complex and imaginary and the system is underdamped with the solution given by:

$$X = E[e^{-(c/2M)t} \sin\left(\sqrt{\frac{K}{M} - \left(\frac{C}{2M}\right)^2}\, t + \psi\right) \tag{21.56}$$

where constants A, B, E and ψ are determined from initial conditions. The graph representing equation (21.56) is shown in Figure 21.11.

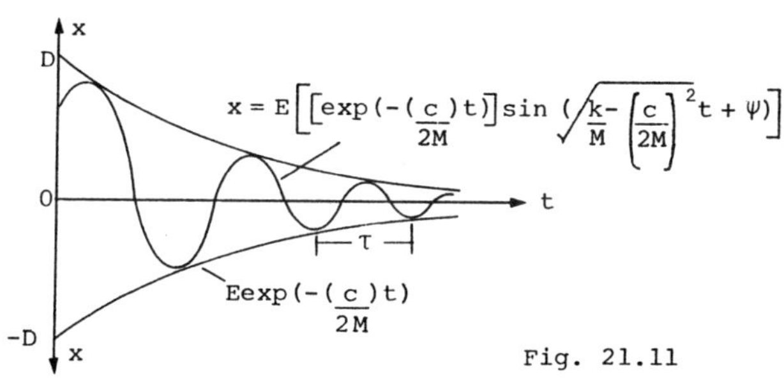

Fig. 21.11

Note: Period of damped vibration 'τ' is given by

$$\tau = \frac{2\pi}{\sqrt{\frac{K}{M} - \left(\frac{c}{2M}\right)^2}} \tag{21.57}$$

21.12 DAMPED FORCED VIBRATION

Damped forced vibrations is the most general case of single degree of freedom vibrating motion.

The equation of motion becomes:

$$M\frac{d^2x}{dt^2} + c\frac{dx}{dt} + kx = F \sin\omega t \tag{21.58}$$

The corresponding spring-mass system is shown in Figure 21.12.

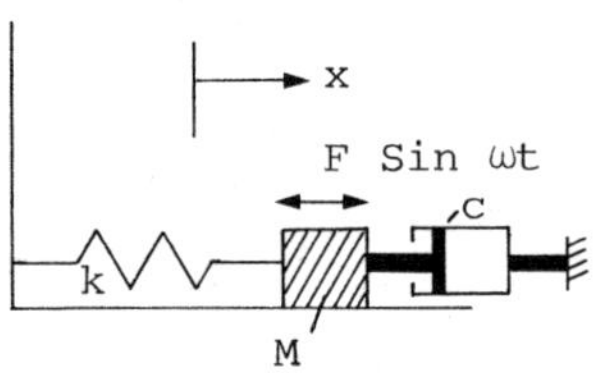

Fig. 21.12

Since equation (21.58) is a second-order, ordinary non-homogeneous differential equation, its solution will consist of the sum of the complementary solution and the particular solution. The homogeneous solution will therefore be identical to equation (21.54), (21.55) or (21.56), and will dampen-out, leaving only the particular or steady-state solution. Assuming the particular solution to be of the form $x_p = A \sin(\omega t - \psi)$, the constants A and ψ are given by

$$\Lambda = \frac{F/K}{\sqrt{\left[1 - \frac{4m^2\omega^2}{C^2_{critical}}\right]^2 + \left[\frac{4m\omega C}{C^2_{critical}}\right]^2}} \qquad (21.59)$$

$$\psi = \tan^{-1}\left[\frac{c\,\omega/k}{\left(1 - \frac{m\,\omega^2}{K}\right)}\right] \qquad (21.60)$$

Note: Angle ψ is the phase difference between the resulting steady-state vibration and the applied force.

The magnification factor is defined as:

$$\text{Magnification factor} = \frac{A}{F/K}$$

$$= \frac{1}{\sqrt{\left[1 - \frac{4m^2\omega^2}{C^2_{critical}}\right]^2 + \left[\frac{4m\omega c}{C^2_{critical}}\right]^2}} \qquad (21.61)$$

and the graph is shown in Fig. 21.13. Resonance occurs only when the damping is zero and the frequency ratio is one.

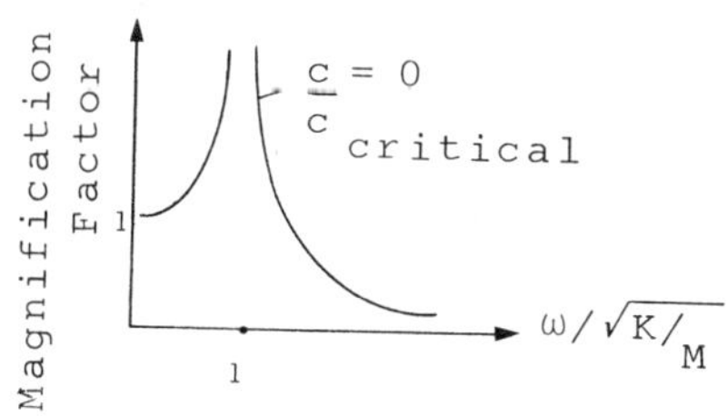

Fig. 21.13

21.13 STABILITY

Taking the equilibirum position of a system as the reference and if a disturbance is imposed, then the system is

A) Neutral if it moves either towards or away from equilibrium.

B) Stable if it comes back to equilibrium, or

C) Unstable if it moves away from equilibrium.

Considering the potential energy v:

If $\frac{dv}{dq} = 0$ at equilibrium then

A) if $\frac{d^2v}{dq^2} > 0$, it is stable

B) if $\frac{d^2v}{dq^2} < 0$, it is unstable and

C) if $\frac{d^2v}{dq^2} = 0$, then higher order terms must be considered.

21.14 POTENTIAL ENERGY FUNCTION

Given a one degree of freedom system, the potential energy function is:

$$v(q) = k_0 + k_1 (q-b) + \frac{1}{2!} k_2(q-b)^2 + \ldots \ldots + \frac{1}{n!} k_n(q-b)^n \tag{21.62}$$

where

$$k_n = \left(\frac{d^n v}{dq^n}\right)_{q=b}$$

If a point of equilibrium is $q = b$, then

$$v(q) = k_0 + \frac{1}{2!} k_2(q-b)^2 + \ldots \tag{21.63}$$

The stability of $q = b$ depends on the first non-zero term after K_0 of the expansion:

A) If n is even then it is stable if $d^n v/dq^n$ is positive.

B) If n is odd or if $d^n v/dq^n$ is negative, then it is unstable.

For a system with n degrees of freedom, the potential energy function is given by

$$v(q_1, q_2, \ldots q_n) = \frac{1}{2}(k_{11}q_1^2 + 2k_{12}q_1q_2 + k_{22}q_2^2 + \ldots) \tag{21.64}$$

where

$$k_{11} = \left(\frac{\partial^2 v}{\partial q_1^2}\right)_{q_1=q_2=\ldots=q_n=0}$$

$$k_{12} = \left(\frac{\partial^2 v}{\partial q_1 \partial q_2}\right)_{q_1=q_2=\ldots=q_n=0}$$

Equation (21.64) is in quadratic form and if it is zero for all q, then the configuration $q_1 = q_2 = \ldots = q_n = 0$ is stable.

21.15 OSCILLATION WITH A SINGLE DEGREE OF FREEDOM

The kinetic energy for a one degree of freedom oscillation is given by

$$T = \frac{1}{2}\mu\dot{q}^2 \tag{21.65}$$

where μ may be expanded as,

$$\mu = \mu(0) + \left(\frac{d\mu}{dq}\right)_{q=0} q + \ldots \tag{21.66}$$

$\mu = \mu(0) = $ constant if $q = 0$ is an equilibrium position.

The Lagrangian Function:

$$L = T - V = \frac{1}{2}\mu\dot{q}^2 - \frac{1}{2}kq^2 \tag{21.67}$$

where $\quad k = k_2 = \left(\frac{d^2v}{dq^2}\right)_{q=0}$

Lagrange's Equation of Motion:

$$\mu\ddot{q} + kq = 0 \tag{21.68}$$

If $q = 0$ is stable:

$$k > 0$$

$$\omega = \sqrt{k/\mu} \quad \text{and}$$

$$q = q_0 \cos(\omega t + \psi) \tag{21.69}$$

where

q_0 = Amplitude and ψ = Phase angle

21.16 GENERAL THEORY OF VIBRATING SYSTEMS

Consider a system with 'n' degrees of freedom.

The kinetic energy function is given by:

$$T = \sum_i \sum_j \frac{1}{2}\mu_{ij}\dot{q}_i\dot{q}_j \tag{21.70}$$

where μ_{ij} is constant and equal to the value at the equilibrium configuration, i.e., with $q_1 = q_2 = \ldots = q_n = 0$.

Potential Energy:

$$v = \sum_i \sum_j \frac{1}{2} K_{ij} q_i q_j \tag{21.71}$$

The Lagrangian Function:

$$L = \sum_i \sum_j \frac{1}{2} (\mu_{ij} \dot{q}_i \dot{q}_j - K_{ij} q_i q_j) \tag{21.72}$$

The Equation of Motion:

$$\sum_i \mu_{ij} \ddot{q}_i = - \sum_i K_{ij} q_i \quad (j=1,2,\ldots) \tag{21.73}$$

In matrix notation:

$$M\ddot{q} = -Kq \tag{21.74}$$

where

$$M = \begin{bmatrix} \mu_{11} & \mu_{12} \cdots \\ \mu_{21} & \mu_{22} \cdots \\ \vdots & \\ \cdots & \cdots \end{bmatrix} = [\mu_{ij}]$$

$$K = \begin{bmatrix} K_{11} & K_{12} \ldots \\ K_{21} & K_{22} \ldots \\ \vdots & \\ \cdots & \cdots \end{bmatrix} = [K_{ij}]$$

The generalized displacement vector:

$$q = \begin{bmatrix} q_1 \\ q_2 \\ \vdots \\ q_n \end{bmatrix}$$

If a solution exists of the form:

$$q_j = B_j \cos(\omega t + \phi_j)$$

then the equation in matrix notation becomes

$$\boxed{(K - M\omega^2)q = 0} \qquad (21.75)$$

with the requirement (for a nontrivial solution) that the determinant:

$$|K_{ij} - \mu_{ij}\omega^2| = 0$$

The roots give the normal frequencies and their corresponding eigenvectors give the normal modes.

By methods of linear algebra:

If M^{-1} exists, then

$$\boxed{(M^{-1}K - I\omega^2)q = 0} \qquad (21.76)$$

where I is the identity matrix, and the equation becomes an eigenvalue problem.

The secular equation is

$$\boxed{\mathrm{Det}(M^{-1}K - I\omega^2) = 0} \qquad (21.77)$$

and the same results are obtained.

21.17 NORMAL COORDINATES

Kinetic Energy:

$$T = \frac{1}{2}\dot{q}^T M\dot{q} \tag{21.78}$$

Potential Energy:

$$v = \frac{1}{2} q^T Kq \tag{21.79}$$

These two may be written in a simpler form when expressed in terms of coordinates such that M and K are diagonal (the secular determinant equation also becomes diagonal). By quating to zero each diagonal term, the roots are determined.

Normal coordinates are determined by first defining the matrix B:

$$q = Bq'$$

$$q' = B^{-1}q$$

Then T and v become

$$T' = \frac{1}{2}\dot{q}'^T B^T MBq' \quad \text{and}$$

$$V' = \frac{1}{2} q'^T B^T KBq'$$

It is required that the matrices be diagonal

$$B^T MB = M' = \begin{bmatrix} \mu_1' & 0 & 0 \ldots\ldots 0 \\ 0 & \mu_2' & 0 \ldots\ldots 0 \\ 0 & 0 & \mu_3' \ldots\ldots 0 \\ \vdots & \vdots & \vdots \\ 0 & 0 & 0 \ldots\ldots \end{bmatrix}$$

and $B^T KB = K' = \begin{bmatrix} K'_1 & 0 & 0 \ldots\ldots & 0 \\ 0 & K'_2 & 0 \ldots\ldots & 0 \\ 0 & 0 & K'_3 \ldots\ldots & 0 \\ \vdots & \vdots & \vdots & \\ 0 & 0 & 0 \ldots\ldots & \end{bmatrix}$

To obtain the required transformation, first find a_K:

$$\begin{aligned} Ma_K &= \mu'_K \, a_K \qquad (K=1,2,\ldots,n) \\ Ka_K &= K'_K \, a_K \end{aligned} \tag{21.80}$$

Subtracting these we get

$$(K - M\lambda_K)a_K = 0 \qquad \lambda_K = K'_K/\mu'_K \tag{21.81}$$

with eigenvalue $\lambda_K = \omega_K^2$

To obtain B: Find the roots of $|K - M\omega^2| = 0$ and then for each ω_K, solve the corresponding $(K-M\omega_k^2)a_k = 0$ to obtain the eigenvector a_k.

Damped Motion: Given that the damping forces are proportional to the velocities

Lagrange's Equation:

$$\frac{d}{dt}\left(\frac{\partial L}{\partial \dot{q}_k}\right) = \frac{\partial L}{\partial q_k} + P_k \tag{21.82}$$

where P_k is the generalized damping force given in tensor notation as

$$P_k = -c_{nk}\,\dot{q}_n \qquad k = 1,2,\ldots,n \tag{21.83}$$

A normal coordinate transformation may be determined in some cases to yield the following governing equation:

$$\mu'_k \, \ddot{q}_k + C'_k \, \dot{q}'_k + K'_k \, q'_k = 0 \tag{21.84}$$

Motion with outside driving force: In normalized coordinates:

$$\mu'_k \ddot{q}'_k + C'_k \dot{q}'_k + K'_k q'_k = R_k e^{i\omega t} \qquad (21.85)$$

21.18 VIBRATION OF A LOADED STRING

Kinetic Energy:

$$T = \frac{M}{2}(\dot{q}_1^{\,2} + \dot{q}_2^{\,2} + \ldots + \dot{q}_n^{\,2}) \qquad (21.86)$$

Potential Energy (for a section of the string)

$$V = \frac{1}{2} K(q_{a+1} - q_a)^2 \qquad (21.87)$$

'a' is any arbitrary particle

and k = Elastic stiffness.

Potential Energy (For the system):

$$V = \frac{K}{2}[q_1^{\,2} + (q_2 - q_1)^2 + (q_n - q_{n-1})^2 \quad + q_n^2] \qquad (21.88)$$

where K = s/h and h is the equilibrium distance between two points.

Lagrangian Function:

$$L = \frac{1}{2} \sum_a [M\dot{q}_a^2 - K(q_{a+1} - q_a)^2] \qquad (21.89)$$

Lagrangian Equation of Motion becomes:

$$M\ddot{q}_a = -K(q_a - q_{a-1}) + K(q_{a+1} - q_a) \tag{21.90}$$

$a = 1,2,\ldots,n$

Using a trial solution, $q_a = b_a e^{i\omega t}$ where b_a = the amplitude of the ath particle we obtain the following recursion formula:

$$-m\omega^2 b_a = K(b_{a-1} - 2b_a + b_{a+1}) \tag{21.91}$$

where the amplitude of the end points of the string correspond to $b_0 = b_{n+1} = 0$

The secular determinant:

$$\begin{vmatrix} -M\omega^2 + 2K & -K & 0 & \cdots & 0 \\ -K & -M\omega^2+2K & -K & \cdots & 0 \\ 0 & -K & -M\omega^2+2K & \cdots & 0 \\ \vdots & \vdots & \vdots & & \vdots \\ 0 & \cdots & 0 & \cdots & -M\omega^2+2K \end{vmatrix} = 0 \tag{21.92}$$

An alternate method consists of substituting $b_a = c\sin(a\phi)$ into the recursion formula resulting in

$$\omega = 2\omega_0 \sin\frac{\phi}{2} \tag{21.93}$$

where $\omega_0 = \left(\frac{K}{M}\right)^{\frac{1}{2}}$

To determine ϕ, let $(n+1)\phi = N\pi$, N = integer

Normal frequencies are given by

$$\omega_N = 2\omega_0 \sin\left(\frac{N\pi}{2n+2}\right) \qquad (21.94)$$

Amplitude

$$b_a = c \sin\left(\frac{N\pi a}{n+1}\right) \qquad (21.95)$$

where $a = 1,\ldots,n$ denotes a particle and $N = 1,2,\ldots,n$ denotes a normal mode.

21.19 WAVES

The typical wave may be written as

$$\frac{\partial^2 q}{\partial t^2} = \nu^2 \frac{\partial^2 q}{\partial y^2} \qquad (21.96)$$

where

ν is the wave speed equal to $\sqrt{\frac{Ka^2}{m}}$ or $\sqrt{s/\rho}$ where $K = s/a$, s = the tension of the string, a = the distance between particles at equilibrium and ρ = the density of the string.

The general solution of equation (21.96) may be expressed as

$$q = \alpha(y+\nu t) + \phi(y-\nu t) \qquad (21.97)$$

Sinusoidal Waves: Given that $q = q(y,t)$ then

$$q = c_1 g\left[\frac{2\pi}{\lambda}(y+\nu t)\right] + c_2 g\left[\frac{2\pi}{\lambda}(y-\nu t)\right] \qquad (21.98)$$

where 'g' represents either sin or cos and 'λ' is the wavelength.

Frequency

$$f = \nu/\lambda = \omega/2\pi$$

Note: Because the wave equation is a linear partial differential equation, lienar combinations of solutions also satisfy the equation.

Standing Waves - The sum of two waves traveling in opposing directions with equal amplitudes results in standing waves which are represented by the following equation:

$$\boxed{q = c \sin\left(\frac{2\pi y}{\lambda}\right) \cos \omega t} \qquad (21.99)$$

where $\omega = 2\pi\nu/\lambda$